THE
WATER
ATLAS

THE
WATER
ATLAS

Robin Clarke
and
Jannet King

THE NEW PRESS

NEW YORK
LONDON

Produced for The New Press by
Myriad Editions Limited
6–7 Old Steine, Brighton BN1 1EJ, UK
www.MyriadEditions.com

Published in the United States by The New Press, New York, 2004
Distributed by W. W. Norton & Company, Inc., New York

ISBN 1-56584-918-3 (hc.)
ISBN 1-56584-907-8 (pbk.)
CIP data is available

The New Press was established in 1990 as a not-for-profit alternative to the large,
commercial publishing houses currently dominating the book publishing industry. The
New Press operates in the public interest rather than for private gain, and is committed
to publishing, in innovative ways, works of educational, cultural, and community value
that are often deemed insufficiently profitable.

The New Press
38 Greene Street
New York, NY 10013
www.thenewpress.com

In the United Kingdom:
6 Salem Road
London W2 4BU

Edited and coordinated for Myriad Editions by Jannet King and Candida Lacey
Design and graphics by Corinne Pearlman and Isabelle Lewis
Cartography by Isabelle Lewis

Printed in Hong Kong through Phoenix Offset Limited
under the supervision of Bob Cassels, The Hanway Press, London

2 4 6 8 10 9 7 5 3 1

CONTENTS

Part 3: Water Health

Part 4: Re-shaping the Natural World

PREFACE

More than four centuries after Mercator helped people to see our world through maps, we now have a way of looking at the Earth, and at ourselves, through what we are doing with water on this planet.

Water is one of the distinguishing characteristics of this planet. Simply put, no life whatsoever is possible without it. Every day we are reminded by missions to Mars and outer space that water is a fundamental sign of life as we know it.

This atlas graphically sounds alarm bells and signals a red alert about the assault and damage we are doing to the planet's life blood: draining wetlands, over-irrigating farmland, contaminating water systems, damming rivers, mining groundwater, tearing down forests, expanding cities, using vast amounts of water for high-tech industry and interfering with the world's climate. Repeated blows are being dealt to a hydrological cycle that has continuously renewed and replenished Earth's water flows since time immemorial.

For the most part, those of us who live in water-rich regions of the world tend to take water for granted. Yet, the worldwide demand for water is increasing every year. By halfway through this century the demand for water in some countries is projected to outstrip supply, relegating nearly half of humanity to conditions of water scarcity.

Lurking at the heart of the struggle over our water future is a clash between two contrasting visions of life itself. On the one hand are those who believe that water is simply an economic good or commodity to be bought and sold in the marketplace for profit, and distributed according to the ability to pay. On the other hand, there are those who believe that water, essential to life, is something to be conserved and made equally available to all people and to nature.

At this moment in time, when everything appears to be up for sale, some of the world's most powerful corporations are lining up to take advantage of lucrative profit opportunities in what is estimated to be a trillion-dollar industry. Backed by global financial and trade institutions, their track record is terrible. In Europe, as well as in major cities throughout the global south, the big water service companies jack up water prices, cut off customers who cannot pay, fail to meet environmental targets, reduce water quality, and engage in bribery and corruption. What is more, as humans pollute, divert and deplete surface water supplies, the race is on to mine precious sources of groundwater. Large water-bottling companies are draining local aquifers, polluting the environment and charging a thousand times more than the same water would cost coming from a public tap.

However, around the world local communities and groups are fighting back. Through a number of community 'water wars', a new social movement is emerging. This citizens' movement is reclaiming water as a 'right' – one that belongs to both people and nature – that should not be bought and sold on the open market. It is characterized by both community-based resistance and community-based alternatives.

Humanity faces a bleak future unless we collectively face up to the coming global water crisis. We know the answers to a water-secure world: water recycling; sustainable agriculture instead of industrial agriculture; massive infrastructure repairs; conservation and reclamation of destroyed water systems; strong laws against pollution; limits on industrial growth; locally appropriate technologies; an end to big dams; severe limits on groundwater extractions. Most importantly, our water-secure future must be based on three foundations: water equity, water conservation, and water democracy.

The building of this community-based water movement may well be the last, best hope for people and nature on this planet. This atlas, which so graphically maps the contours of life-giving water – its realities and threats and connectedness in our daily lives – will surely become an important tool in the building of this social movement and the survival of the planet. May it also become a challenge to every reader to become an active part of the solution.

Tony Clarke, Polaris Institute
Maude Barlow, Council of Canadians
co-authors of *Blue Gold: The Fight to Stop the Corporate Theft of the World's Water*

INTRODUCTION

In the first few days of the year 2004, a deal was imminent between Turkey and Israel – a deal that says much about the planet Earth's hydrological and military plight: 'water for tanks', we might call it. Under the agreement, Israel agreed to export arms to Turkey in exchange for fresh water, to be delivered along the east coast of the Mediterranean by tanker.

Tankers are more normally associated with the liquid gold known as oil. Iraq's northern neighbour may have little oil but it is exceptionally well placed to trade in water, of which it has significant amounts compared to its mostly parched neighbours.

The deal will involve Israel building a fleet of giant water tankers to ship 50 million cubic metres of water a year for 20 years from the river Manavgat in Anatolia, and Turkey buying Israeli tanks and air force technology. The amount of water to be imported amounts to 3 percent of Israel's current needs. For its part, Turkey plans on becoming a freshwater superstar, using this precious resource to bring in hundreds of millions of dollars a year in hard currency. Turkey already delivers water by tanker to Cyprus and now plans to sell Manavgat water to Malta, Crete and Jordan.

Israel, of course, is a country with particular hydrological, military and industrial problems. It uses far more water per capita than any other state in the region. It is already desperately short of water and in the next 20 years expects to need much more for farming and industrial development. Yet many of its water reserves are in danger from salt water intrusion, and the levels of its reservoirs and major freshwater lakes often fall to dangerously low levels. It is as well that the country is relatively rich: the cost of importing water from Turkey is estimated at more than one dollar per cubic metre.

Moving water from country to country by tanker is not going to solve the world's water problems, though it may ease the situation in a few local situations. The need for water on this planet cannot be over-emphasized. Everyone needs it and those who don't get it die. It is the substance used to characterize Earth as 'The Blue Planet'. It is water that made life here possible; conversely, a lack of water would make life impossible. Indeed, a lack of water does make life impossible in many places on Earth and a lack of water often causes the mass migration of peoples. The forced evacuation of the Marsh Arabs from their desiccated homeland in the Tigris/Euphrates valley is just the latest and most tragic example.

Water is thus the very stuff of life and should be the intergalactic symbol for our planet. Unlike other resources such as forests and soil, which can be both destroyed and rejuvenated, water is a fixed quantity resource. The planet houses approximately 1,386 million cubic kilometres of it in its seas, lakes, rivers, aquifers, ice, snow and water vapour. Every drop counts because that drop never disappears or reproduces, even though it may change its state from liquid to gas to solid and back to liquid, and travel from the Antarctic to the Sahara and then on to the Russian steppes. Global warming looks as though it will accelerate all this, increasing rates of evaporation and hence of precipitation. In most places, warmer will also mean wetter in terms of increased rainfall but drier in terms of increased evaporation rates.

The conventional literature will tell you that water has become a global issue because of two things: more and more people need it, and each needs more of it as living standards rise. At best these are half-truths, but half-truths hide multitudes of sins and the world of water has become a mucky business on this planet, in more ways than one.

The muck that we throw into our water is the first issue. Industry used to be the worst offender but it has improved its act in many countries. Agriculture has not and relies still on the principle of dilution to somehow rid us of noxious pesticides, herbicides, fertilizers and animal manure. Town and city managers are mostly not much better; they like to wash their urban sewage into water courses outside their domain where they do not have to worry about it and quietly bury solid wastes while hoping the toxic materials they contain will never contaminate underground aquifers. Their hopes are not always fulfilled but new legislation on the recycling of wastes, notably in the European Union, promises to effect marked improvements.

In effect, water pollution reduces the volume of water available for use by human and other populations. Much of our wastewater is discharged into the hydrological system without purification. Since a cubic metre of contaminated wastewater spoils up to 10 cubic metres of pure water, our waste disposal habits have been calculated to consume the equivalent of twice the annual discharge of the world's greatest river, the Amazon, or about one-third of the world's annual run-off from the land to the sea.

The details implied by such broad-brush statistics are not to be entirely trusted because hydrological date are notoriously unreliable and difficult to obtain (*see page 93*). However, there can be little doubt that water pollution threatens a substantial portion of the planet's renewable water. As much is confirmed by practical observation: people on many continents are

11

condemned to use dirty water to fulfil their daily needs for washing, cooking and drinking. Thus there is already not enough water to go round given our present habits of conspicuous consumption and pollution.

This one fact accounts for much of the ill health and many of the early deaths that occur in developing countries. According to the World Health Organization, an estimated 1.7 million deaths are caused annually by dirty water. Most of these deaths are of children and are the result of dehydration from diarrhoea caused by ingesting faecal bacteria. Many more people are crippled by water-related disease, unable to work or contribute to the household economy. It is estimated that 82 million years of healthy life are lost annually in developing countries as a result of dirty water.

Again, the detailed statistics cannot be trusted. These figures are certainly very conservative since they include only known, reported deaths. A 1997 United Nations report claimed that water-related diseases caused between 5 and 10 million deaths a year, and that one-half of all those living in the developing world suffer at any given moment from a disease caused by drinking contaminated water or eating contaminated food. Perhaps. But again the broad-brush picture is certainly accurate enough: dirty water causes immense suffering and high rates of premature death throughout the developing countries. Recent calculations suggest that between 2000 and 2020 almost as many people may die from water-related diseases as from AIDS.

However, people in developing countries do not always demand ever greater quantities of water as their lifestyles improve. For one thing, in many areas, their lifestyles are not improving. For another, what they want is not more water but cleaner water. If you have to trek several kilometres through the bush to fetch a daily water supply, or pay up to US$3 per cubic metre to buy water from a passing tanker lorry, you do not squander what you have or make huge efforts to get more. What is needed is cleaner water made more readily and cheaply available.

It seems that we have turned a blind eye for many decades to a host of water-connected issues. A massive increase in big dam construction during the last decades of the past century forced an estimated 80 million people to move home. They protested but no one took much notice. Their valleys and their homesteads were flooded anyway. In the United States the destruction of the Everglades wetlands caused concern but 'only' from environmentalists. Elsewhere, the draining of wetlands continued as part of a deliberate policy to rid the world of 'unhealthy swamps'. With the swamps went rare species of wildlife and nature's unique system of natural cleansing. Under heavy rains, the new concrete wetlands flood even more quickly than the natural ones, but with much greater inconvenience for the inhabitants.

Most of the water used in the world is for irrigation. Much of that water never actually reaches the crop it is supposed to support. Nevertheless, the repeated irrigation of the same land often results in a build-up of salts and, eventually, the salinization of good farming land. Million of hectares have been rendered useless in this way, and much of the need for new irrigation schemes is caused by waterlogging and salinization in older ones.

Irrigation is also an effective method of distributing agricultural chemicals to places where they do maximum harm. One result is that the Mississippi river in the USA has become so polluted with agricultural chemicals washed in from the American prairies that an area the size of New Jersey in the Gulf of Mexico into which the Mississippi discharges has become an ecologically dead zone incapable of supporting any form of life. Even that is not irrigation's best claim to infamy which surely rests with the death of the Aral Sea in central Asia. Here the diversion of incoming water to irrigate cotton and other crops has reduced the sea to a fraction of its former size, demoting what were once fishing ports to the status of inland ghost towns, depriving thousands of fishermen of their jobs and visiting disease upon those without sufficient resources to move away.

In a water-short world, mistrust and insecurity are what mark most relations between countries that share rivers. There are many of them: more than 260 river basins are international and 13 are shared by five or more countries. In West Asia, nations scrap over the waters of the Yamak, the Euphrates and the Tigris, and there is serious conflict over water between Israel and Palestine; Mexico and the USA argue about the Colorado and the Rio Grande; Egypt, Ethiopia and the Sudan all want more of the Nile waters; and Bangladesh, Bhutan, India and Nepal have frequent disputes about the Brahmaputra and the Ganges.

Not that such disputes cannot be resolved peacefully. They can. A complicated system of treaties governs how the waters of the Mekong are to be used, and many others govern the Rhine's flow through Europe, for example. However, it may well be significant that such agreements appear to be more easily reached in water-rich Europe and Southeast Asia than in countries where flows and supplies are periodic and limited.

12

Commercial conflict is another matter but equally serious. Unsavoury accounts of soft drink manufacturers robbing local farmers of their water resources are not hard to find in the literature. Fierce conflicts rage over limited water resources between farms and cities and industries in many countries.

To all this there is an easy and fashionable answer. Let the market sort it out. Make water a commodity, like tin, fish, rail transport and even education, and all will be well. In this way, human needs can be met. Multinational companies have responded joyously to this call to arms, taking over from government the irksome business of managing and distributing the world's water resources, while making huge profits for themselves and their directors. They have been energetically supported by the purveyors of bottled water, selling a basic resource which does not belong to them in the first place for extortionist amounts.

There are, of course, good arguments for not subsidising the extravagant use of water by large-scale farmers, golf clubs and recreational facilities. There are even better arguments for believing that access to reasonable quantities of clean water is not just a human need but a human right – as is the right to life, to food, to health. By implication, these rights cannot be achieved without access to clean water. More specifically the first major UN conference on water, held in Mar del Plata in 1977, stated that "All peoples, whatever their stage of social development and their social and economic conditions, have the right to have access to drinking water in quantities and of a quality equal to their basic needs."

If access to adequate water is a basic human right, it will not necessarily be met by market forces. States therefore have a duty to ensure that their populations' rights are fulfilled. They have a long way to go. And it is not unreasonable for them to ask which direction to take.

During the 1960s, a group of far-sighted people coined the term 'soft technology', defining this as 'technology that is valid for all people for all time'. They thus pre-empted today's concept of sustainable development, arguing that we should use only those technologies that were not harmful to people who did not have them and which would not reduce options for future generations.

In 1976 the US physicist Amory Lovins pursued these concepts in some detail in relation to energy, describing a future in which soft energy paths would lead to a durable peace. He foresaw the use of soft technologies to fulfil diminishing energy needs, with reductions in the use of fossil and nuclear fuels. We are at the beginning of just such a path into the future.

The route to a water future of the kind described by Ibrahim Diaw *(see page 91)* will surely follow similar paths. It will not involve such 20th-century concepts as giant dams, reversing the flow of rivers and nuclear desalination plants. It will involve small-scale water harvesting, drip irrigation, clever conservation techniques, improving information on when to irrigate, allocating specific water resources to the environment and a host of other small-scale measures. Above all, it will involve reductions in water use in many areas. No one wants to consume water; they want to grow food, manufacture the goods they need and live comfortably. If these goals can be met using much less water – and they can – we will be on the way to a water-secure world. The pioneer of soft paths to this future is the US engineer Peter H. Gleick. As he says:

> The soft path for water strives to improve the productivity of water use rather than seek endless sources of new supply. It delivers water services and qualities matched to users' needs, rather than just delivering quantities of water. It applies economic tools such as markets and pricing, but with the goal of encouraging efficient use, equitable distribution of the resource, and sustainable system operation over time. And it includes local communities in decisions about water management, allocation, and use.

This publication is best seen as a tool to speed our journey along this road.

ACKNOWLEDGEMENTS

The publishers are grateful to following organizations for permission to reproduce their photographs, and for their assistance in the production of this book:

Food and Agriculture Organisation (FAO)
Photographic Library
<www1.fao.org/media_user/_home.html>
5, 18 FAO/17283/J. Holmes; 5, 28 & 90 FAO/16964/U. Keren; 37 FAO/18030/I. Balderi; 6, 46, 91 FAO/19705/G. Bizzarri; 6, 74 FAO/13940/M. Marzot; 7, 82 FAO/13702/J. Isaac; 90 FAO/10070/J. Van Acker; 7, 92 FAO/13463/I. de Borhegyi

International Rivers Network (IRN)
<www.irn.org/wcd/gallery.shtml>
45 Pat Morrow

US Army Corps of Engineers Digital Visual Library
<images.usace.army.mil/main.html>
6, 59 Adrien Lamarre

WaterAid
89 John Spaull

World Food Programme
<http://www.wfp.org>
49 Evelyn Hockstein

The publishers are grateful to the following organizations for permission to base the map on pages 64–65 on *Groundwater Resources of the World*: Bundesanstalt für Geowissenschaften und Rohstoffe (BGR); Commission for the Geological Map of the World (CGMW); International Association of Hydrgeologists (IAH), UNESCO

The publishers are grateful to Dr Caroline Sullivan of the Centre for Ecology & Hydrology (CEH) Wallingford for permission to represent the Water Poverty Index on pages 88–89.

Whilst every reasonable effort has been made to contact the holders of copyrigh materials used in the atlas, the authors and publisher will gladly receive information that will enable them to rectify any inadvertent errors or ommissions in subsequent editions.

ABOUT THE AUTHORS

Robin Clarke is the author of many books on the environment including *Water: The International Crisis, The Science of War and Peace, We All Fall Down, The Challenge of the Primitives* and *Science and Technology in World Development.* He was editor of UNEP's flagship publications *Global Environment Outlook 2000* and *Global Environment Outlook 2002* as well as its *GEMS Environment Library.* He is the editor of the World Meterological Organization's *World Climate News.*

Jannet King has worked for many years editing and researching environmental, political, and historical atlases, including those on Endangered Species, Food, and the State of the World, also produced by Myriad Editions.

PART 1

A FINITE RESOURCE

MAKE NO MISTAKE: THE WORLD WATER SUPPLY IS IN CRISIS, and things are getting worse, not better. In spite of the many grandiose plans made by the United Nations and other international bodies since the 1970s, the basic issues have yet to be tackled in practical terms. The situation will continue to worsen until effective action is taken on a worldwide basis.

The basic problem is that the volume of fresh water on the Earth's surface is fixed: it can be neither increased nor decreased. It follows, then, that as populations grow, and the aspirations of individuals increase, less and less water is available per person. In water-rich countries, such as Canada and Gabon, this matters little; in some dry areas of water-rich countries, such as the south-west of the USA, the situation locally is already alarming, with cities, farms and industries fighting over the control of limited resources; in much of the rest of the world, many people already face water penury. By 2050, more than 4 billion people – nearly half the world population – are expected to live in countries that are chronically short of water.

Being chronically short of water means many things: that there is not enough water to grow crops and support livestock; the need for long, and at least daily, journeys to carry water from stand-pipes, ponds or springs; having to wash in, and sometimes drink, dirty water that others have recently washed in or drunk; having dirty clothes, dirty dishes and being thirsty; and, people with little money having to spend too much of their tiny incomes on water vendors. Water shortage is a major barrier to development and an important reason why so many of the world's poor are still poor. And, as shown in Part 3, this is one of the main reasons why so many people in affected countries are ill or die young.

Water shortage conjures up images of rich people irrigating their lawns, taking long showers and over-filling their swimming pools. Of course, such habits do not help, but statistically they are a drop in the ocean. More than two-thirds of all the water withdrawn worldwide is used to water crops and animals, the vast majority of it being used to irrigate crops in arid and

Twice as much water was used worldwide in 2000 as in 1960

semi-arid areas. Here, waste is really high, with only a small percentage of the water that is used ever reaching the crop for which it was intended.

Industry is the second biggest user – 21 percent of the world total. In only a few highly industrialized countries, such as the USA, the Netherlands and Germany, is more water consumed by industry than by agriculture. But although industrial demands are limited in terms of quantity, much of what industry withdraws is consumed – that is, it is so polluted that it cannot be easily re-used. In comparison, the amount of water used for domestic purposes, including municipal use in towns and cities, is relatively trivial – about 10 percent of the total.

Much of our water comes from underground aquifers. These vast water reserves are critical for the survival of human populations, and contain more than 100 times as much as all the planet's surface water reserves put together. But aquifers are renewed only slowly, by rainwater percolating down through soil and rock. Today, we are quite literally robbing the aquifer bank, extracting water at rates far in excess of natural renewal.

There are some immediate consequences. Water tables are retreating and wells are drying up. Water pumps have to work harder and harder to bring to the surface the water we need. As water levels fall, so does the ground above. Severe ground subsidence is a common feature of cities, from Beijing to Mexico City, that depend largely on aquifers for their supplies. The use of such water to support urban populations is entirely understandable; the use of it to irrigate crops – in particular wheat – in arid countries is not. Such food crops can be easily purchased from water-rich countries, which is in effect an efficient way of importing water from wet to dry areas. Saving aquifers for uses where there are no other alternatives makes good common sense.

One further use has not yet been mentioned: leaving water to run its natural course to fuel the world's rivers, lakes and wetlands. This vital use has often been ignored. Yet the health of the planet depends on just this use of water, and we ignore it at our peril.

The surprising thing about the world's water is that its volume never changes.

The planet is always home to approximately 1,386 million km³ of water. Nearly all of this (97.5 percent) is salt water, contained in the oceans, seas, saltwater lakes and salty aquifers (underground reserves). Of the 2.5 percent that is fresh water, more than two-thirds is unavailable for human use because it is locked up in glaciers, snow, ice and permafrost.

Of the fresh water that is technically 'available' for people to use, only a tiny proportion is found on the surface of the Earth – in lakes, rivers, wetlands, the soil, air humidity, and in plants and animals. All the rest is stored in underground aquifers. Although this groundwater is a key resource in many countries, it is being used faster than it is replenished.

Surface water is in constant motion. It is evaporated from the land and oceans by the sun's heat, which turns liquid water into water vapour. In the atmosphere this vapour then condenses to form the water droplets of which clouds are formed.

The key to our survival is that some of the water that evaporates from the oceans falls on land, feeding the rivers, watering the soil, and restocking the underground aquifers. This is the renewable part of our freshwater supplies, on which we all depend.

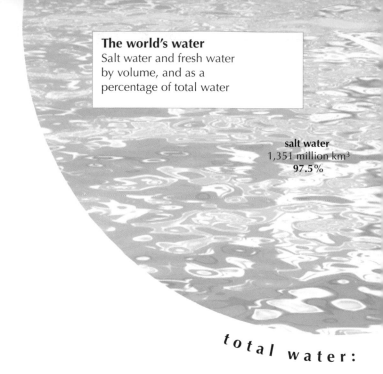

The world's water
Salt water and fresh water by volume, and as a percentage of total water

salt water
1,351 million km³
97.5%

total water:

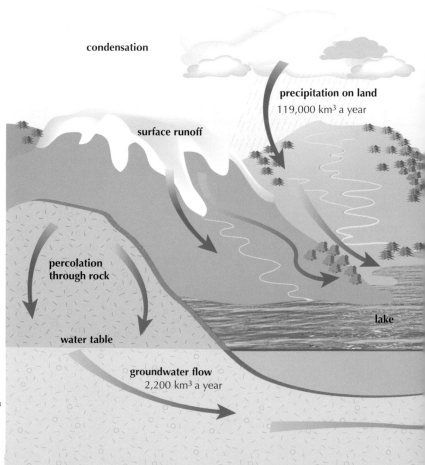

condensation

precipitation on land
119,000 km³ a year

surface runoff

percolation
through rock

water table

lake

groundwater flow
2,200 km³ a year

Freshwater sources
By volume, and as a percentage of total fresh water

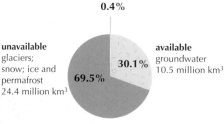

available
lakes; soil moisture; air humidity; marshes and wetlands; rivers; plants and animals
135,000 km³
0.4%

unavailable
glaciers; snow; ice and permafrost
24.4 million km³

69.5%

30.1%

available
groundwater
10.5 million km³

FRESH OUT OF WATER

**Only 2.5% of the world's water is fresh.
More than two-thirds of this is
unavailable for human use.**

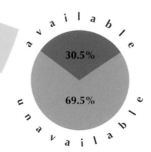

fresh water
35 million km³
2.5%

1,386 million km³

available
30.5%

69.5%

unavailable

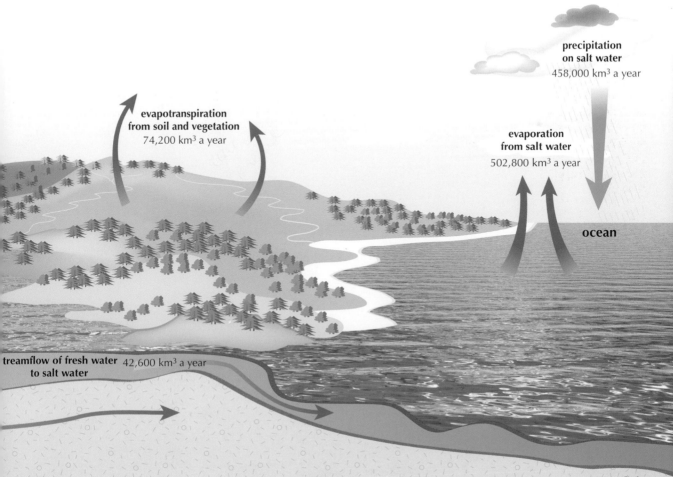

**precipitation
on salt water**
458,000 km³ a year

**evapotranspiration
from soil and vegetation**
74,200 km³ a year

**evaporation
from salt water**
502,800 km³ a year

ocean

treamflow of fresh water 42,600 km³ a year
to salt water

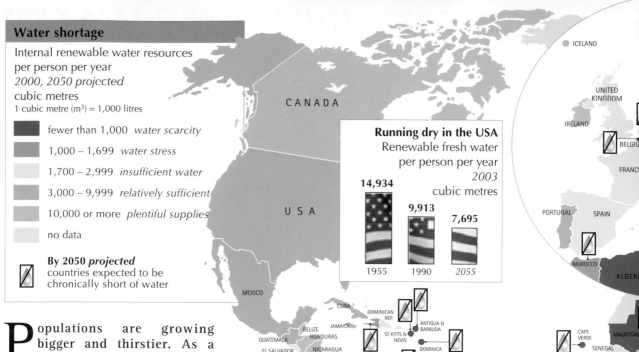

Water shortage

Internal renewable water resources
per person per year
2000, 2050 projected
cubic metres
1 cubic metre (m³) = 1,000 litres

- fewer than 1,000 *water scarcity*
- 1,000 – 1,699 *water stress*
- 1,700 – 2,999 *insufficient water*
- 3,000 – 9,999 *relatively sufficient*
- 10,000 or more *plentiful supplies*
- no data

By 2050 *projected*
countries expected to be
chronically short of water

Running dry in the USA
Renewable fresh water
per person per year
2003
cubic metres

14,934 — 1955
9,913 — 1990
7,695 — *2055*

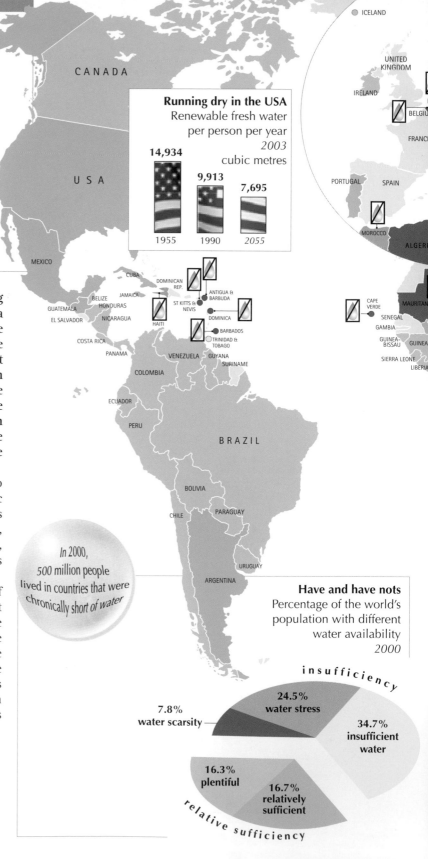

Populations are growing bigger and thirstier. As a result, 500 million people are living in countries where water is chronically in short supply, and a further 2.4 billion are living in countries where the water system is under stress. The situation is likely to worsen, with populations projected to increase in many of the countries that are already short of water.

Enough rain falls on land each year to provide, on average, nearly 7,000 cubic metres of fresh water per person. This is more than enough for most needs, but the water is not evenly distributed, and people are not free to move to areas of abundant water.

Countries in the driest areas of Africa and Asia are among the most water-deprived in the world, and the way in which governments manage the increasing water crisis will be crucial. With each state claiming the right to the water flowing through its territory, countries downstream are in danger of finding their supplies drying up.

*In 2000,
500 million people
lived in countries that were
chronically short of water*

Have and have nots
Percentage of the world's
population with different
water availability
2000

insufficiency

24.5% water stress
34.7% insufficient water
7.8% water scarcity
16.3% plentiful
16.7% relatively sufficient

relative sufficiency

MORE PEOPLE, LESS WATER

More than one-third of the world's population is short of water, and the situation is getting worse.

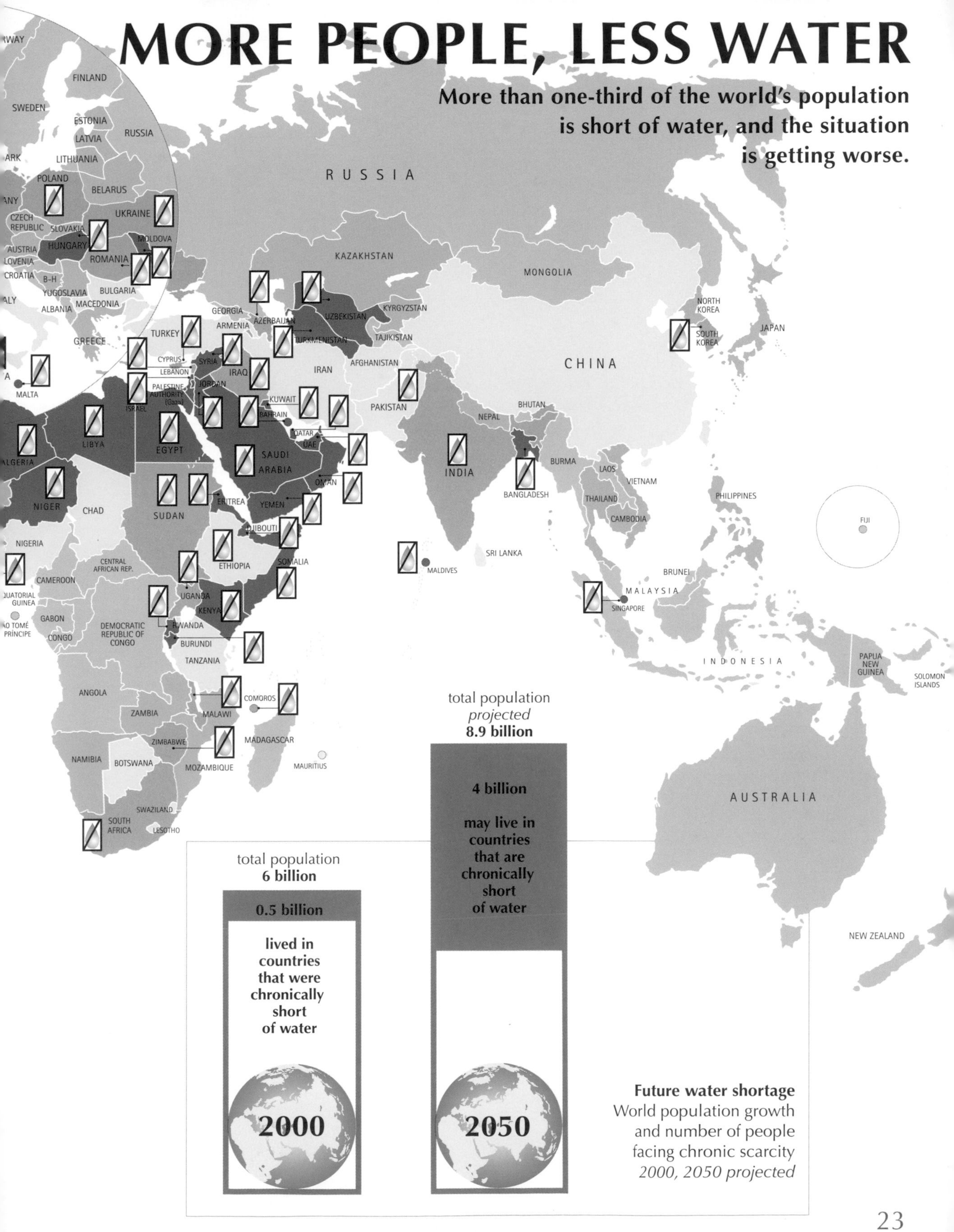

total population
projected
8.9 billion

4 billion

may live in
countries
that are
chronically
short
of water

total population
6 billion

0.5 billion

lived in
countries
that were
chronically
short
of water

2000

2050

Future water shortage
World population growth
and number of people
facing chronic scarcity
2000, 2050 projected

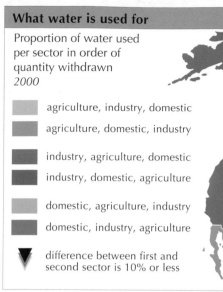

Nearly 4,000 cubic kilometres of fresh water are withdrawn every year – an average of around 1,700 litres per person per day. But while the total amount of fresh water in the world remains the same, the amount of water withdrawn per person is increasing.

Most water is withdrawn for use in agriculture, especially in the drier regions of the world. In Europe and North America, industry predominates, with power generation using the most.

By comparison, the water people drink, or use to keep themselves, their clothes, dishes and houses clean, is relatively insignificant. Worldwide, domestic withdrawal amounts to an average of about 170 litres per person per day. However, this figure is kept artificially low by the difficulty many people in the developing world have in obtaining water for domestic purposes (*see page 30*).

Much of the water that is withdrawn is returned to surface and underground water sources largely unchanged. However, some water is so badly polluted in the process of being used, that it is no longer suitable for human consumption.

World water use
by sector
2000

10% domestic

21% industry

69% agriculture

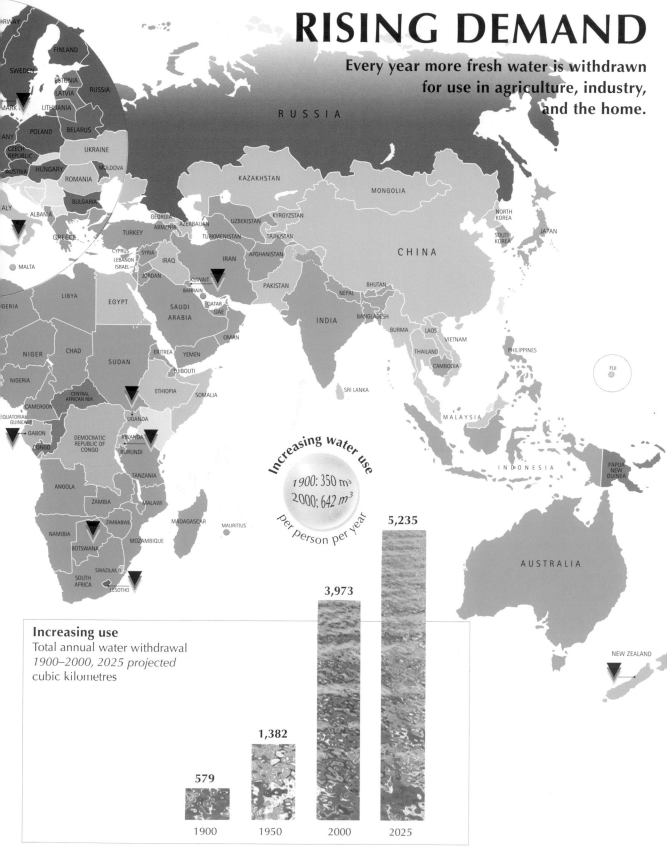

RISING DEMAND

Every year more fresh water is withdrawn for use in agriculture, industry, and the home.

Increasing water use

1900: 350 m³
2000: 642 m³

per person per year

Increasing use
Total annual water withdrawal
1900–2000, 2025 projected
cubic kilometres

579	1,382	3,973	5,235
1900	1950	2000	2025

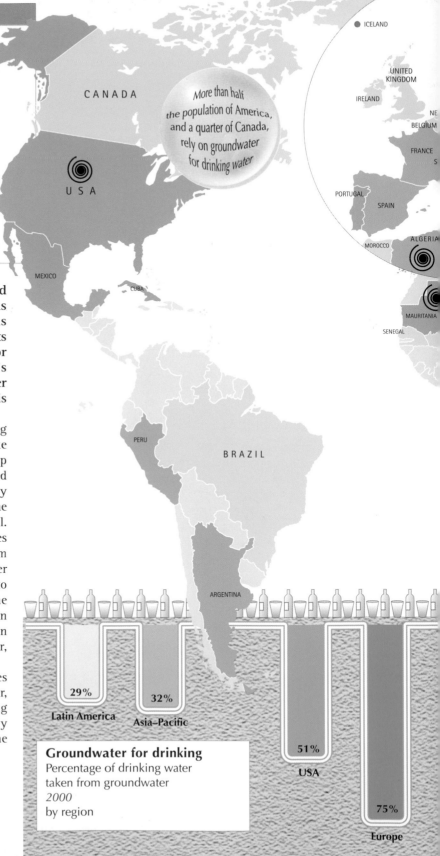

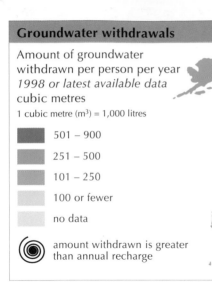

Groundwater withdrawals

Amount of groundwater
withdrawn per person per year
1998 or latest available data
cubic metres

1 cubic metre (m³) = 1,000 litres

- 501 – 900
- 251 – 500
- 101 – 250
- 100 or fewer
- no data

amount withdrawn is greater
than annual recharge

> *More than half the population of America, and a quarter of Canada, rely on groundwater for drinking water*

Vast stores of water are held beneath the ground in areas of porous rock known as aquifers. This groundwater represents the only source of drinking water for about a quarter of the world's population. But in many places water is being withdrawn faster than it is being replaced.

Even where aquifers are being replenished, because it may take centuries or millennia for water to seep back down through the rock, the replaced water may not be available for many generations. In some regions, such as the Sahara, aquifers are not recharged at all.

Much of the world's agriculture relies on irrigation systems that use water from aquifers. Although more efficient water pumps have enabled food production to increase, they are now leading to the over-use of groundwater. Already, in northern China, restrictions have been placed on the use of groundwater, causing a drop in the grain harvest.

Many of the world's largest cities depend almost entirely on groundwater, but the rate at which water is being extracted to cope with such densely populated areas is not sustainable. One day, the wells will simply dry up.

Groundwater for drinking
Percentage of drinking water
taken from groundwater
2000
by region

29% Latin America
32% Asia–Pacific
51% USA
75% Europe

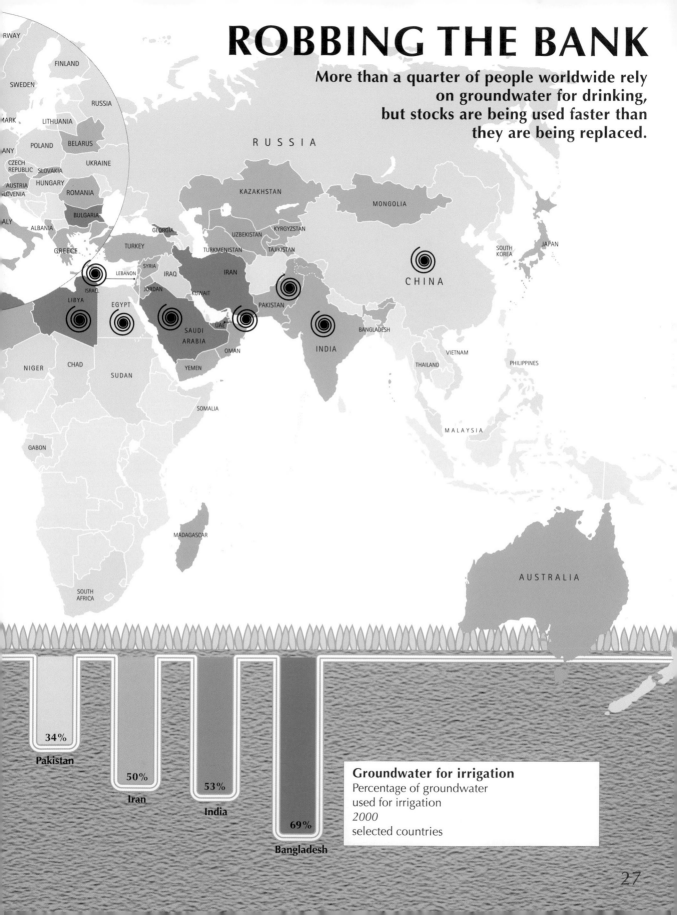

ROBBING THE BANK

More than a quarter of people worldwide rely on groundwater for drinking, but stocks are being used faster than they are being replaced.

Groundwater for irrigation
Percentage of groundwater used for irrigation
2000
selected countries

34% Pakistan

50% Iran

53% India

69% Bangladesh

PART 2

USES AND ABUSES

"YOU CANNOT FILL THE ARAL WITH TEARS," wrote the young Uzbek poet Mukhammed Salikh more than 20 years ago. He was lamenting the death of the great inland Aral Sea: the conversion of a 66,000-square-kilometre lake that used to support 60,000 fishermen and produce 40,000 tonnes of fish a year into a poisoned wasteland.

Twenty years ago few had even heard of the greatest environmental disaster ever created by humans. Today, the story of the diversion of the water that once fed the Aral Sea – some 50 cubic kilometres a year – for irrigation, reducing inflow to zero, is well known. The sea is now half its former size, water levels have fallen by more than 13 metres, and the mineral content has increased fourfold, effectively killing off the fish population. Commercial fishing ended in 1982 and what were once seaside towns and villages, now littered with the wrecks of former fishing boats, find themselves 70 km from the coastline which has been heavily polluted by pesticides and heavy metals. Half the populations of some Aral cities have simply fled; those that remain battle with a deadly mix of pollutants that cause cancer, bronchitis, kidney and liver diseases, and arthritis. Contaminated water has led to a 30-fold increase in the incidence of typhoid. Infant mortality in the Aral region is among the world's highest.

The death of the Aral Sea is the most dramatic example of the abuse of water. But it is far from unique. Its converse, the flooding of vast valleys to produce reservoirs for hydropower, irrigation and flood control, has been almost as costly. Worldwide, as many as 80 million people may have been forced to flee their homes by such schemes – people who have lost not only their houses but also their livelihoods and ways of life. Reservoirs have much to answer for: the loss of floodplains, cultural heritage and wildlife also come high on the list.

Water pollution has wreaked enormous damage. In many developing countries, the rivers below major towns and cities have been turned into what are, in effect, open sewers. Industries have been permitted to spew out rich concoctions of toxic chemicals without regard for downstream fish, animals and humans. Surface and underground water supplies have become contaminated by both organic pollutants and the nutrients used by agriculture. Irrigation has turned millions of hectares into waterlogged, salinized land that will be hard, even impossible, to recover. Deforestation has bared hillsides, leading to more frequent and more serious flashfloods downstream. In industrialized countries, tap water has now to be so intensively treated that the chlorine it contains often makes it undrinkable.

While not every cloud has a silver lining, some still do. Irrigation, the biggest single user of the world's water, has helped to feed billions, and continues to do so. The hydropower provided by reservoirs supplies one-fifth of the world's electricity, without increasing atmospheric levels of greenhouse gases, or causing the intractable problem of how to dispose of dangerous radioactive materials. The water used by industry has helped millions escape the drudgery of agricultural labour and raised standards of living in developing countries.

Farming is responsible for 70% of water pollution in the USA

And yet, the cost of such progress has been too high, far too high: polluted water, scarcer water resources, floodplains turned to concrete, wildernesses destroyed. The happy balance, better known these days as sustainable development, is not much in evidence although there certainly has been some progress: a slowing in the rate of dam construction and the development of irrigation, an increase in small-scale hydropower construction, attempts at water conservation in the home, the field and the factory, a new emphasis on creating 'more crop per drop', a tendency to eat less water-intensive foods, the increased use of wastewater in irrigation.

So far these are not much more than just a drop in our polluted freshwater ocean. But they are a beginning. The major ways in which these trends, and others, might be developed into a sound overall plan for the rational management of water on this planet are summarized in Part 6.

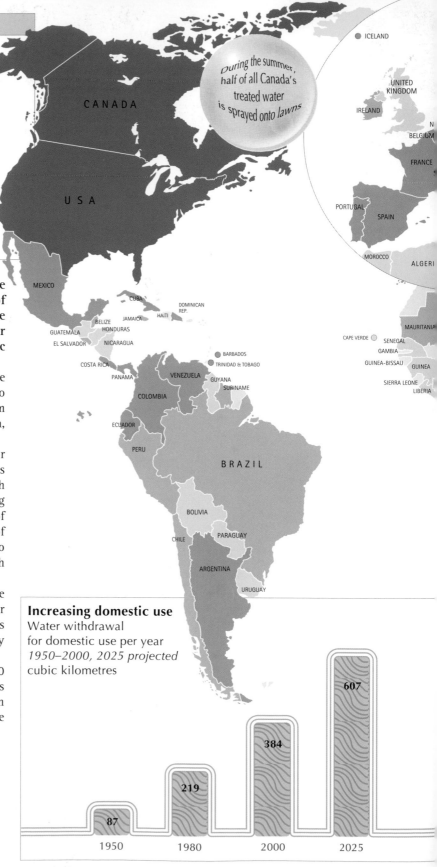

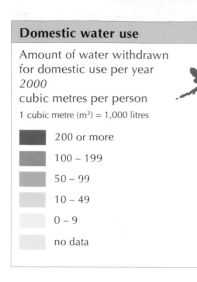
During the summer, half of all Canada's treated water is sprayed onto lawns

Water usage is one of the most conspicuous forms of consumption. As people become richer and enjoy a higher standard of living, so their domestic water use increases.

The amount of water used in the home, or by municipal authorities to supply residential areas, varies from more than 800 litres per day in Canada, to only one litre in Ethiopia.

Much of the water withdrawn for domestic purposes never even reaches the consumer, but is lost through leaking pipes. Cities in developing countries typically lose 40 percent of their water through leakage. Some of this water finds its way back to groundwater, rivers or lakes, but much of it evaporates.

Leaking taps in homes may waste more water than is actually used for cooking and drinking. And as much as 30 percent of domestic water is simply flushed down the toilet.

In some developing countries, 20 litres of water per person per day is considered luxurious. Some people in developed countries use much more than this just to water their lawns.

Increasing domestic use

Water withdrawal
for domestic use per year
1950–2000, 2025 projected
cubic kilometres

Year	Value
1950	87
1980	219
2000	384
2025	607

WATER AT HOME

Only 10% of all water withdrawn is intended for domestic purposes, but the amount used varies widely between countries.

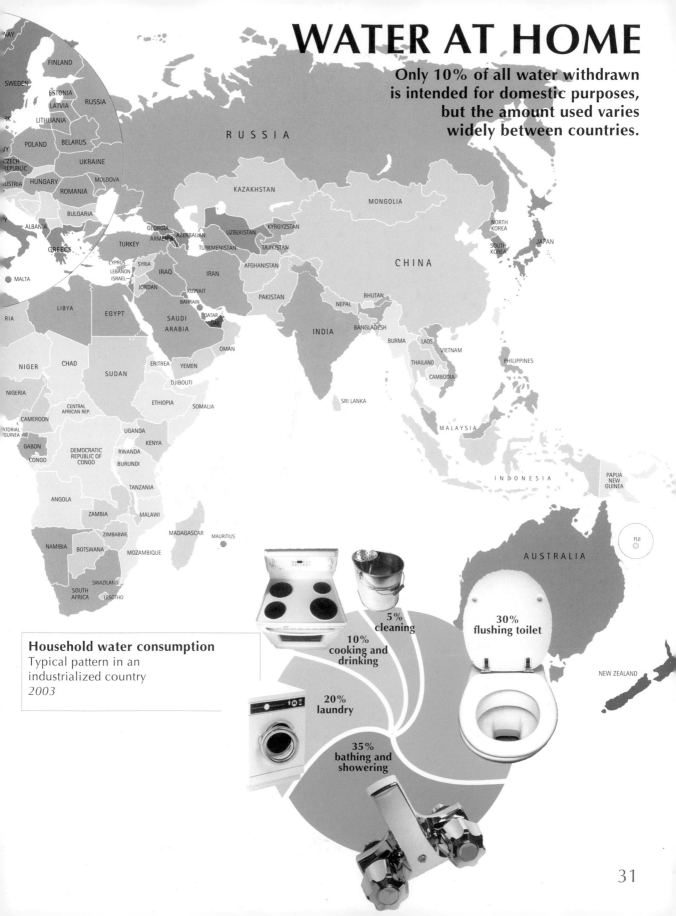

Household water consumption

Typical pattern in an industrialized country
2003

5% cleaning

10% cooking and drinking

30% flushing toilet

20% laundry

35% bathing and showering

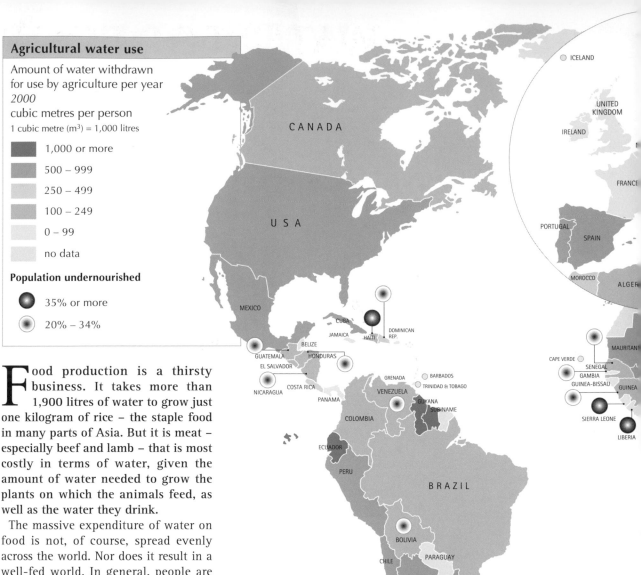

Agricultural water use

Amount of water withdrawn
for use by agriculture per year
2000
cubic metres per person

1 cubic metre (m³) = 1,000 litres

- 1,000 or more
- 500 – 999
- 250 – 499
- 100 – 249
- 0 – 99
- no data

Population undernourished

- 35% or more
- 20% – 34%

Food production is a thirsty business. It takes more than 1,900 litres of water to grow just one kilogram of rice – the staple food in many parts of Asia. But it is meat – especially beef and lamb – that is most costly in terms of water, given the amount of water needed to grow the plants on which the animals feed, as well as the water they drink.

The massive expenditure of water on food is not, of course, spread evenly across the world. Nor does it result in a well-fed world. In general, people are better fed in wet regions than in dry, but the strain of providing enough water for agriculture puts enormous stress on the environment.

One way of resolving the problem of the global water shortage is to grow more food using less water – getting 'more crop per drop'.

Food as water
tomatoes: 95% water
apples: 85% water
hot dogs: 56% water

WATER FOR FOOD

Nearly 70% of all freshwater withdrawals are used for agriculture, yet millions of people remain malnourished.

Water for food

Minimum amount of water
needed to produce
1 kilogram of food
2000

1 kg potatoes	500 litres
1 kg wheat	900 litres
1 kg sorghum	1,100 litres
1 kg soybeans	1,650 litres
1 kg rice	1,900 litres
1 kg poultry	3,500 litres
1 kg beef	15,000 litres

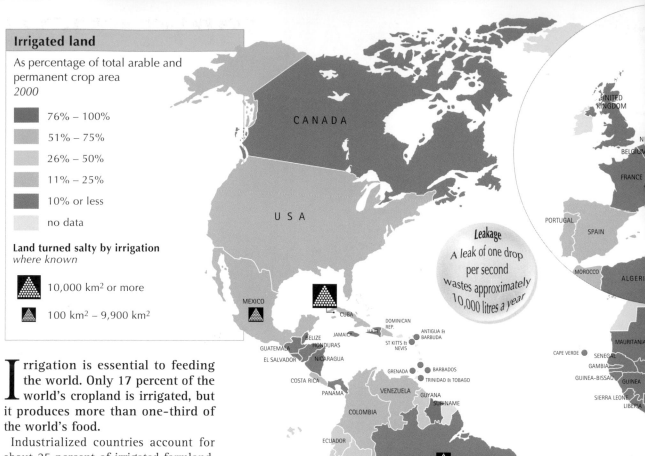

Irrigated land

As percentage of total arable and permanent crop area
2000

- 76% – 100%
- 51% – 75%
- 26% – 50%
- 11% – 25%
- 10% or less
- no data

Land turned salty by irrigation
where known

- 10,000 km² or more
- 100 km² – 9,900 km²

Leakage
A leak of one drop per second wastes approximately 10,000 litres a year

I rrigation is essential to feeding the world. Only 17 percent of the world's cropland is irrigated, but it produces more than one-third of the world's food.

Industrialized countries account for about 25 percent of irrigated farmland. But the rate at which irrigation is being introduced is beginning to slow down because of a lack both of suitable land and water supplies, and the high capital cost – up to US$10,000 per hectare.

Many developing countries are using up to 40 percent of their renewable fresh water for irrigation. But more than half of this is lost through leakage and during distribution, so never reaches the crops.

If irrigated fields are not properly drained, they can become waterlogged. Salts build up in the soil, making the land infertile. This problem has affected about 30 percent of irrigated land.

The key to improved irrigation lies in the more efficient use of water, in the recycling of wastewater, and in better drainage (*see page 86*). Several countries now use treated wastewater for irrigation; Israel was using up to 30 percent of its urban wastewater in this way as early as 1987.

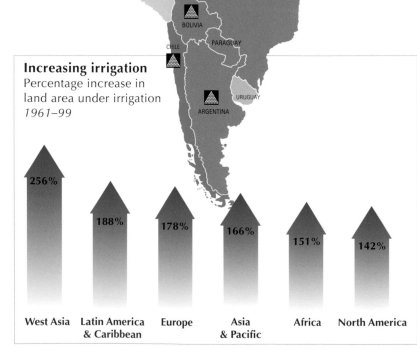

Increasing irrigation
Percentage increase in land area under irrigation
1961–99

West Asia	Latin America & Caribbean	Europe	Asia & Pacific	Africa	North America
256%	188%	178%	166%	151%	142%

IRRIGATION

Irrigated land is usually more productive than non-irrigated land, but poorly managed irrigation can lead to water-logged or infertile soil.

Salinity
This is estimated *to* affect 23% of irrigated land in China and 21% in Pakistan

The River Nile never reaches the sea. Its waters are used so extensively to irrigate Egypt's crops that at times the flow into the Mediterranean is reduced to nothing.

Decline of the Aral Sea
1957, 1984, 2001

- water
- dry land

The Aral Sea in Kazakhstan has shrunk by 50% in area and by over 66% in volume since 1957, largely because of the diversion of the Amu Darya and Syr Darya rivers for irrigation (see page 61).

Aral Sea
1957

Aral Sea
1984

Aral Sea
2001

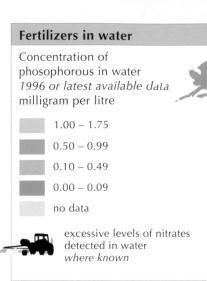
Abbotsford-Sumas aquifer
*Nitrate levels found to exceed US
and Canadian drinking water
health standards*

East Anglia
*Fertilizer from intensive
farming has led to
over 50 mg/litre of
nitrates in 142 locations*

Nebraska and Kansas
*More than 45 mg/litre of
nitrates in 35%
of samples*

Canary Islands, Spain
*Use of nitrates on banana
plantations has polluted
wells, with 70 – 265 mg/litre
being detected*

Yucatán Peninsula
*Agricultural run-off and
human waste has led to
levels of more than
45 mg/litre of nitrates
in shallow groundwater*

Animal slurry
Farm animals in the USA
produce 130 times more waste
than humans do,
much of which finds its way
into the water system

A griculture is becoming increasingly industrialized. The chemicals that are used in the process run off into rivers and lakes, leach into the soil, and contaminate the water we drink.

Fertilizer use in the industrialized world soared between 1960 and 1980. It subsequently declined in Europe, but in the developing world, where fertilizers offer a quick-fix solution to chronic food shortages, their use is increasing. Many countries are now experiencing problems as a result.

Phosphates and nitrates are spread on soil to promote growth, but can have a disastrous effect on freshwater lakes, where they lead to an abundance of oxygen-hungry algae and weeds, which deprive fish and other aquatic life of the oxygen they need to survive. Nitrates in water supplies are becoming a hazard for animals and humans (*see page 56*).

When water rich in nitrates is used to irrigate crops that are also being fertilized, it can reduce their yield, and make them more vulnerable to pests and diseases. This, in turn, leads to an increase in pesticide use. The most notorious, DDT, has been banned in many countries, but is still present in water systems worldwide.

AGRICULTURAL POLLUTION

Farming on an industrial scale, and the use of chemicals to boost yields, is building up problems for the world's water supplies.

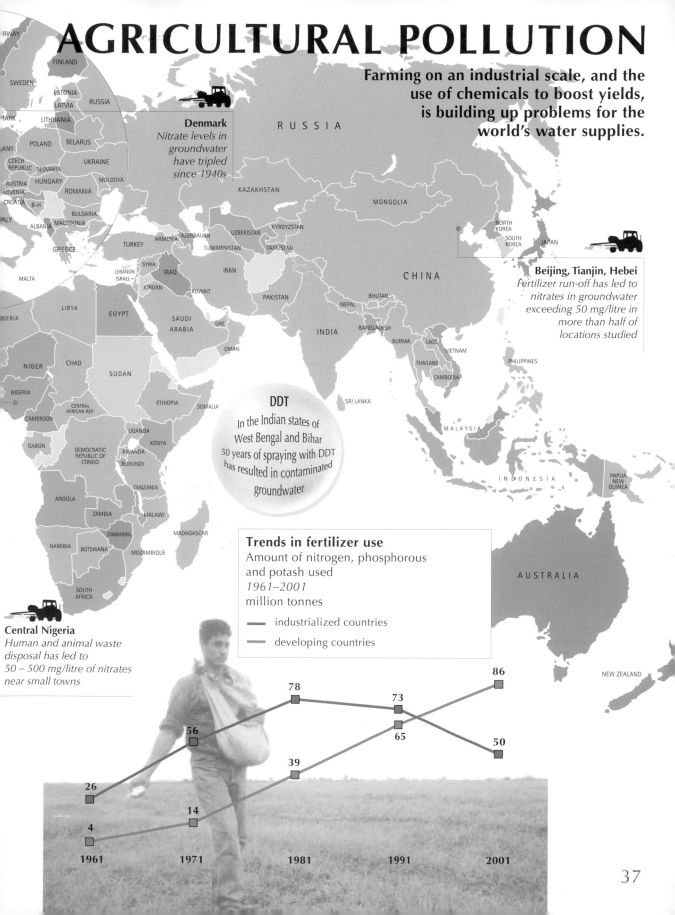

Denmark
Nitrate levels in groundwater have tripled since 1940s

Beijing, Tianjin, Hebei
Fertilizer run-off has led to nitrates in groundwater exceeding 50 mg/litre in more than half of locations studied

DDT
In the Indian states of West Bengal and Bihar 50 years of spraying with DDT has resulted in contaminated groundwater

Central Nigeria
Human and animal waste disposal has led to 50 – 500 mg/litre of nitrates near small towns

Trends in fertilizer use
Amount of nitrogen, phosphorous and potash used
1961–2001
million tonnes

— industrialized countries
— developing countries

	1961	1971	1981	1991	2001
industrialized	26	56	78	73	50
developing	4	14	39	65	86

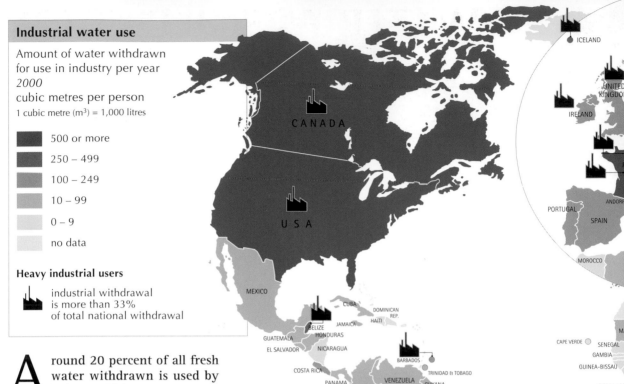

Industrial water use

Amount of water withdrawn
for use in industry per year
2000
cubic metres per person
1 cubic metre (m³) = 1,000 litres

- 500 or more
- 250 – 499
- 100 – 249
- 10 – 99
- 0 – 9
- no data

Heavy industrial users

industrial withdrawal
is more than 33%
of total national withdrawal

Around 20 percent of all fresh water withdrawn is used by industry. This amounts to an average of about 130 cubic metres per person per year, although more than half of this is used by hydropower plants, and for cooling in power stations, with much of the the water being returned to its source virtually unchanged.

Other major industrial users of water are much heavier consumers. They include chemical and petroleum plants, metal industries, the wood, pulp and paper industry, food processing, and machinery manufacture.

In high-income countries 59 percent of all water withdrawn is used by industry, although since 1980 the industrial use of water has increased only slowly as a result of concerted efforts to control it. However, the industrial use of water around the world is expected to rise steeply over the next 25 years as more countries industrialize.

The fear is that this industrialization will add to the growing problem of water pollution. In developing countries, 70 percent of industrial waste is dumped untreated into water, where it pollutes both underground and surface water supplies.

Increasing industrial use
Water withdrawn
per year
for industrial use
1950–2000, 2025 projected
cubic kilometres

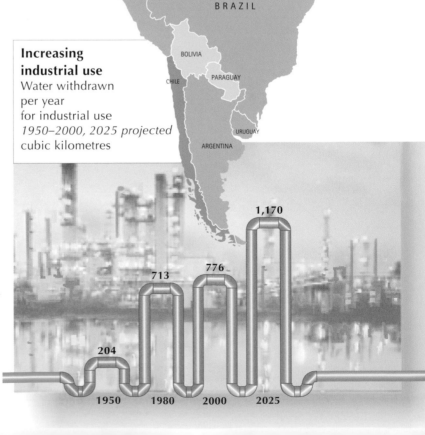

1950	1980	2000	2025
204	713	776	1,170

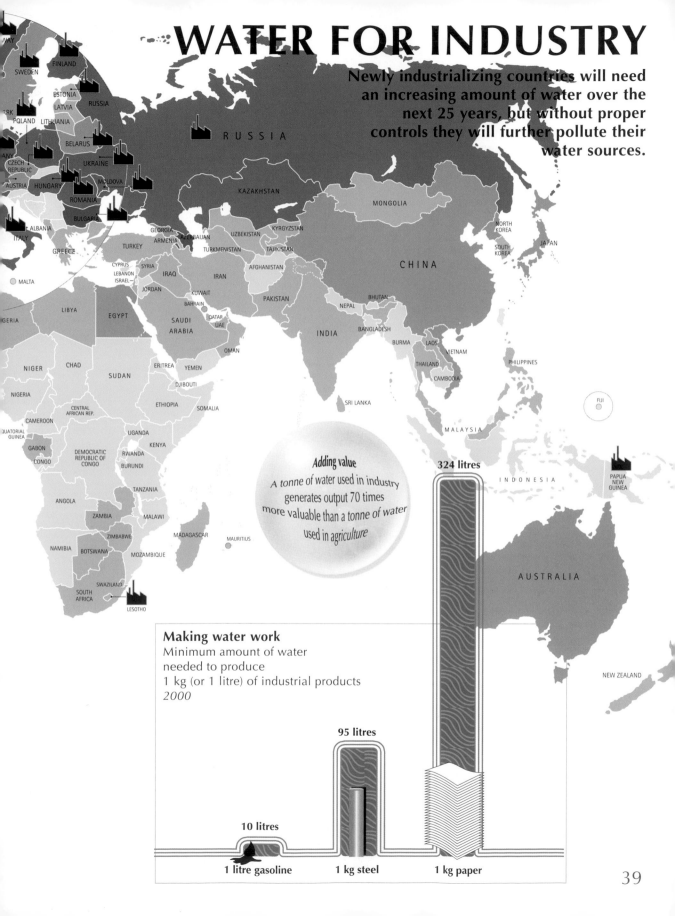

WATER FOR INDUSTRY

Newly industrializing countries will need an increasing amount of water over the next 25 years, but without proper controls they will further pollute their water sources.

Adding value

A tonne of water used in industry generates output 70 times more valuable than a tonne of water used in agriculture

Making water work

Minimum amount of water needed to produce 1 kg (or 1 litre) of industrial products *2000*

324 litres

95 litres

10 litres

1 litre gasoline   1 kg steel   1 kg paper

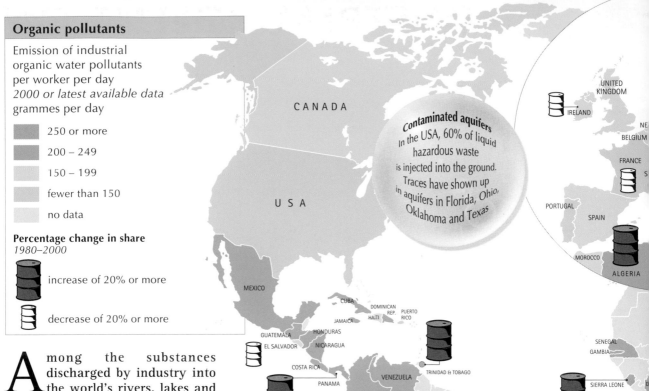

Organic pollutants

Emission of industrial
organic water pollutants
per worker per day
2000 or latest available data
grammes per day

- 250 or more
- 200 – 249
- 150 – 199
- fewer than 150
- no data

Percentage change in share
1980–2000

increase of 20% or more

decrease of 20% or more

Contaminated aquifers
In the USA, 60% of liquid hazardous waste is injected into the ground. Traces have shown up in aquifers in Florida, Ohio, Oklahoma and Texas

Industrial sludge
300–500 million tonnes of heavy metals, solvents, toxic sludge and other wastes accumulate in water sources each year as a result of industrial processes

Among the substances discharged by industry into the world's rivers, lakes and aquifers are organic pollutants that use up vital oxygen in the water, heavy metals such as lead, cadmium and mercury, and some of the most dangerous chemicals ever created – persistent organic pollutants (POPs).

POPs are a group of carbon-based chemicals that share a number of characteristics: they all persist in the environment for a long time, concentrate in the food chain, travel long distances and are linked to serious health effects. They include DDT, and PCBs used for electrical insulation.

Many chemicals are disposed of onto, under, or into the ground. Over time they may travel through the soil and into underground aquifers. The leakage of petrochemicals from tanks at filling stations into aquifers is a growing problem.

Industrial accidents create localized crises, such as occurred in 1996 in the Philippines, when 1.5 million cubic metres of leaked industrial mine waste killed a 27-km long river, the main source of livelihood for local people, and raised the level of zinc in drinking water to 14 times over the safe levels.

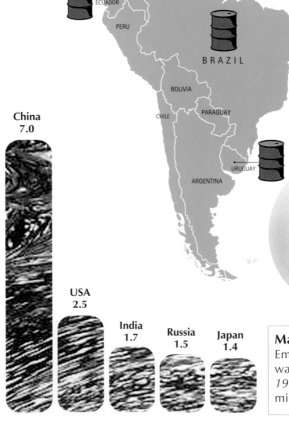

China 7.0
USA 2.5
India 1.7
Russia 1.5
Japan 1.4

Major polluters
Emissions of organic water pollutants per day
1998
million kilogrammes

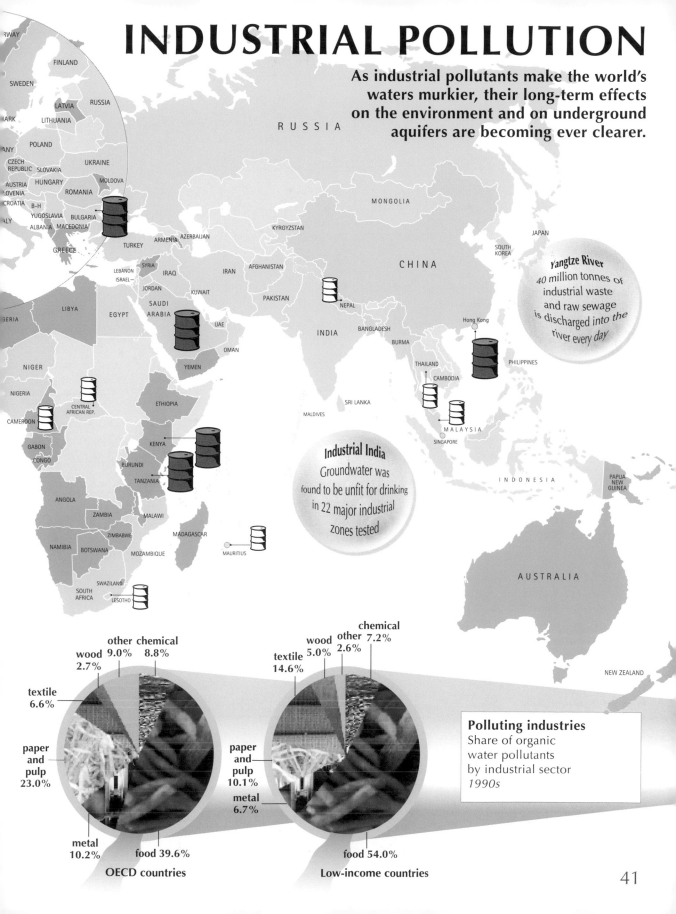

INDUSTRIAL POLLUTION

As industrial pollutants make the world's waters murkier, their long-term effects on the environment and on underground aquifers are becoming ever clearer.

Yangtze River
40 million tonnes of industrial waste and raw sewage is discharged into the river every day

Industrial India
Groundwater was found to be unfit for drinking in 22 major industrial zones tested

Polluting industries
Share of organic water pollutants by industrial sector
1990s

OECD countries

- wood 2.7%
- other 9.0%
- chemical 8.8%
- textile 6.6%
- paper and pulp 23.0%
- metal 10.2%
- food 39.6%

Low-income countries

- wood 5.0%
- other 2.6%
- chemical 7.2%
- textile 14.6%
- paper and pulp 10.1%
- metal 6.7%
- food 54.0%

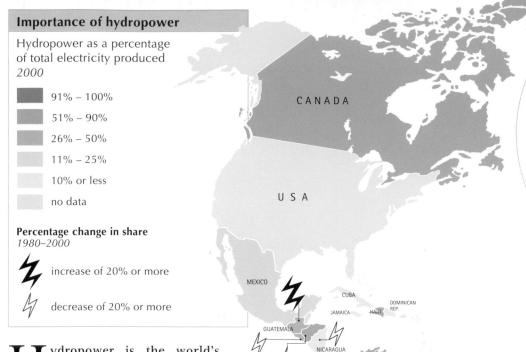

Importance of hydropower

Hydropower as a percentage
of total electricity produced
2000

- 91% – 100%
- 51% – 90%
- 26% – 50%
- 11% – 25%
- 10% or less
- no data

Percentage change in share
1980–2000

increase of 20% or more

decrease of 20% or more

Hydropower is the world's most important source of renewable energy. It produces neither greenhouse gases nor the pollutants associated with burning fossil fuels, and tends to be more acceptable to people than nuclear energy.

The potential for hydropower has not yet been fully realized, with only about a third of possible sites being exploited. Although, in theory, hydropower production could double or even treble, the environmental and social costs of large dams are likely to prevent this.

But there is still potential for small hydropower schemes, which generate less than 10 megawatts – sufficient energy to power a small town. They are ideal for remote rural areas, where the cost of connection to the national grid can be prohibitive.

The reservoirs needed for large hydropower plants are very wasteful of renewable water resources. Worldwide, their total surface amounts to some 500,000 square kilometres – roughly twice the size of the UK, and larger than the state of California. This leads to an enormous loss of water through evaporation before the water has served any useful purpose.

Hydropower in China

More than 45,000 small-scale hydropower plants are benefiting about 300 million people in China. They have provided electricity for the first time in many rural areas. The power of local streams and rivers can be harnessed and turned into electricity by technology that fits into a small hut.

small-scale hydropower is 40% of total output from hydropower

This technology has many ecological advantages. In Sichuan, an estimated 40 million m³ of timber has been saved each year since 1.48 million people have been supplied with electricity.

WATER FOR POWER

Hydropower is the world's most important source of renewable energy and provides nearly one-fifth of the world's electricity.

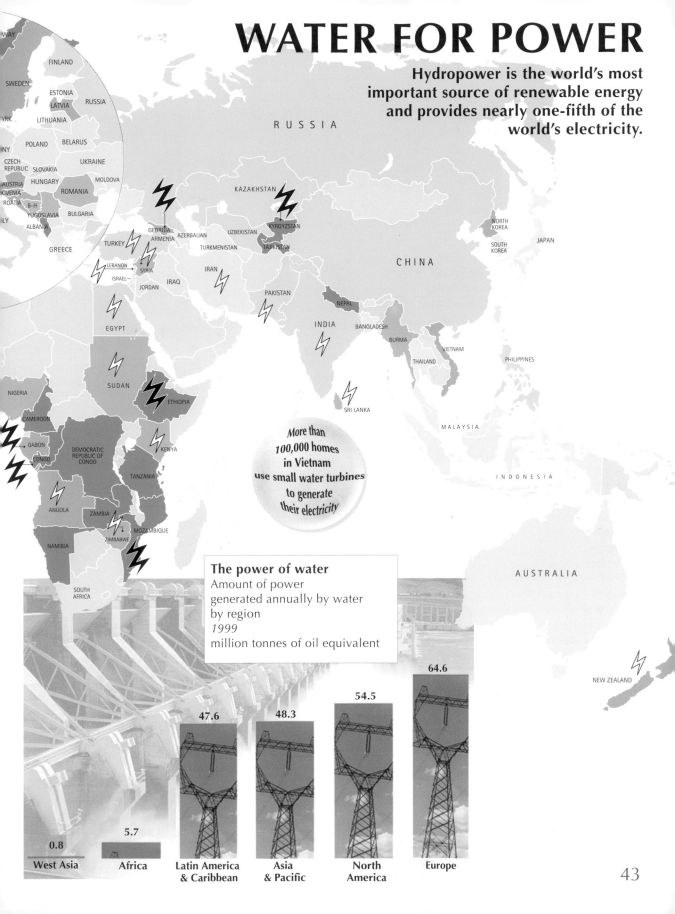

More than 100,000 homes in Vietnam use small water turbines to generate their electricity

The power of water
Amount of power generated annually by water by region
1999
million tonnes of oil equivalent

West Asia	0.8
Africa	5.7
Latin America & Caribbean	47.6
Asia & Pacific	48.3
North America	54.5
Europe	64.6

D ams harness water resources for food production, energy generation, flood control and domestic use. During the 1990s, between US$32 and US$46 billion was spent annually on large dams. Up to 40 percent of irrigated land now relies on dams, and hydroelectric plants generate 19 percent of the world's electricity.

But at what cost? An estimated 80 million people have been deprived of their homes and forced to move elsewhere by dam construction. Many were never consulted, and almost all those who were lost their case. In many areas, dam construction has played havoc with delicate ecosystems, destroying fisheries and eliminating wildlife. In many countries, both natural and cultural resources have been submerged by new reservoirs.

Environmental and human rights groups have campaigned to halt and restrict dam construction. Recent campaigns have centred on the Three Gorges Dam project in China (see box), the Pak Mun in Thailand, Ilisu in Turkey, Ralco in Chile, Epupa in Namibia, the Lesotho Highlands Water Project, and Nam Theun II in Laos.

Partly because of these pressures, the rate of large dam construction has declined since the 1970s. In the USA, the decommissioning rate for large dams has overtaken the rate of construction. In 2000, the World Commission on Dams questioned the value of many dams in meeting water and energy development needs, and paved the way for a different approach to future water development.

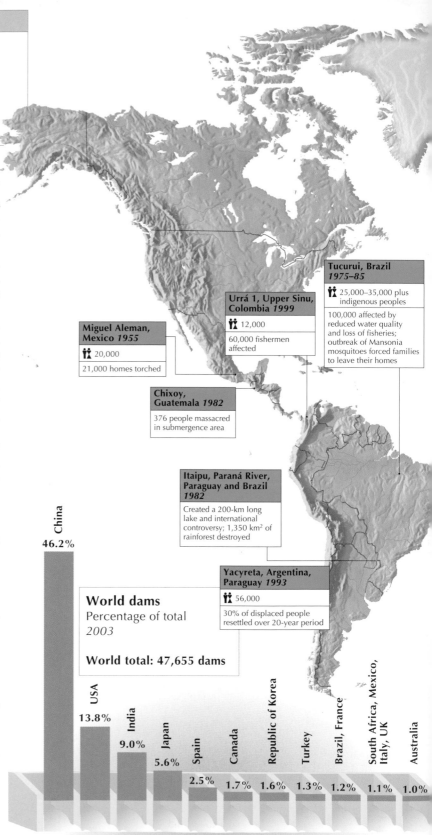

Miguel Aleman, Mexico 1955

👥 20,000

21,000 homes torched

Urrá 1, Upper Sinu, Colombia 1999

👥 12,000

60,000 fishermen affected

Tucurui, Brazil 1975–85

👥 25,000–35,000 plus indigenous peoples

100,000 affected by reduced water quality and loss of fisheries; outbreak of Mansonia mosquitoes forced families to leave their homes

Chixoy, Guatemala 1982

376 people massacred in submergence area

Itaipu, Paraná River, Paraguay and Brazil 1982

Created a 200-km long lake and international controversy; 1,350 km² of rainforest destroyed

Yacyreta, Argentina, Paraguay 1993

👥 56,000

30% of displaced people resettled over 20-year period

World dams
Percentage of total
2003

World total: 47,655 dams

China 46.2%

USA 13.8%

India 9.0%

Japan 5.6%

Spain 2.5%

Canada 1.7%

Republic of Korea 1.6%

Turkey 1.3%

Brazil, France 1.2%

South Africa, Mexico, Italy, UK 1.1%

Australia 1.0%

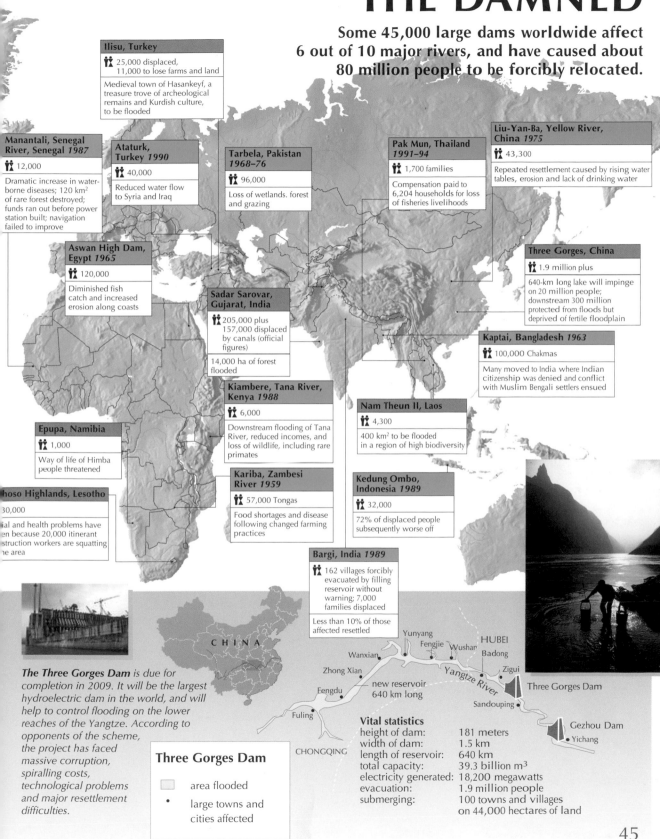

THE DAMNED

Some 45,000 large dams worldwide affect 6 out of 10 major rivers, and have caused about 80 million people to be forcibly relocated.

Ilisu, Turkey

👥 25,000 displaced, 11,000 to lose farms and land

Medieval town of Hasankeyf, a treasure trove of archeological remains and Kurdish culture, to be flooded

Manantali, Senegal River, Senegal 1987

👥 12,000

Dramatic increase in water-borne diseases; 120 km² of rare forest destroyed; funds ran out before power station built; navigation failed to improve

Ataturk, Turkey 1990

👥 40,000

Reduced water flow to Syria and Iraq

Tarbela, Pakistan 1968–76

👥 96,000

Loss of wetlands. forest and grazing

Pak Mun, Thailand 1991–94

👥 1,700 families

Compensation paid to 6,204 households for loss of fisheries livelihoods

Liu-Yan-Ba, Yellow River, China 1975

👥 43,300

Repeated resettlement caused by rising water tables, erosion and lack of drinking water

Aswan High Dam, Egypt 1965

👥 120,000

Diminished fish catch and increased erosion along coasts

Sadar Sarovar, Gujarat, India

👥 205,000 plus 157,000 displaced by canals (official figures)

14,000 ha of forest flooded

Three Gorges, China

👥 1.9 million plus

640-km long lake will impinge on 20 million people; downstream 300 million protected from floods but deprived of fertile floodplain

Kaptai, Bangladesh 1963

👥 100,000 Chakmas

Many moved to India where Indian citizenship was denied and conflict with Muslim Bengali settlers ensued

Kiambere, Tana River, Kenya 1988

👥 6,000

Downstream flooding of Tana River, reduced incomes, and loss of wildlife, including rare primates

Nam Theun II, Laos

👥 4,300

400 km² to be flooded in a region of high biodiversity

Epupa, Namibia

👥 1,000

Way of life of Himba people threatened

Kariba, Zambesi River 1959

👥 57,000 Tongas

Food shortages and disease following changed farming practices

Kedung Ombo, Indonesia 1989

👥 32,000

72% of displaced people subsequently worse off

...hoso Highlands, Lesotho

...0,000

...ial and health problems have ...en because 20,000 itinerant ...struction workers are squatting ...e area

Bargi, India 1989

👥 162 villages forcibly evacuated by filling reservoir without warning; 7,000 families displaced

Less than 10% of those affected resettled

The Three Gorges Dam is due for completion in 2009. It will be the largest hydroelectric dam in the world, and will help to control flooding on the lower reaches of the Yangtze. According to opponents of the scheme, the project has faced massive corruption, spiralling costs, technological problems and major resettlement difficulties.

CHINA

CHONGQING

Yunyang
Fengjie
Wushan
HUBEI
Badong
Wanxian
Zhong Xian
Zigui
Fengdu
new reservoir 640 km long
Yangtze River
Three Gorges Dam
Sandouping
Gezhou Dam
Fuling
Yichang

Three Gorges Dam

☐ area flooded

• large towns and cities affected

Vital statistics

height of dam:	181 meters
width of dam:	1.5 km
length of reservoir:	640 km
total capacity:	39.3 billion m³
electricity generated:	18,200 megawatts
evacuation:	1.9 million people
submerging:	100 towns and villages on 44,000 hectares of land

PART 3

WATER HEALTH

THE WORLD OF THE 21ST CENTURY has much wrong with it. Perhaps the greatest scandal of all is that more than a billion people do not have easy access to a safe supply of fresh water. Many of those that do, do not have even a cold water tap in their house. These facts are made even more depressing by the existence of serious and determined efforts to improve the situation over recent decades. Even our best efforts, it seems, have been doomed to relative failure.

Not having easy access to water means real hardship for millions of people, mostly women. Even easy access is really a misnomer. In the most recent survey of progress in this area, 'access to an improved water supply' is defined as being a household connection or access to a public standpipe, a borehole, a protected dug well, a protected spring or a source of rainwater collection. Only the first would be generally considered 'handy'. All the others may be either at the end of the street or a kilometre away, involving a long walk and heavy burdens. Nor does the definition imply anything about the quality or the reliability of the 'improved' source. Protected wells may not be sufficiently protected, and as everyone in developing countries knows, wells can, and do, dry up.

2.3 billion people suffer from diseases linked to water

The safe disposal of human waste is in an equally lamentable state. More than 2 billion people do not have access to 'improved sanitation' – toilet facilities that are private or shared (but not public), and which prevent contact with faeces by people and other animals, including insects. Equally serious is the fate of even those wastes that are disposed of through improved sanitation facilities – which include pit latrines. Much, if not most, is simply flushed away, without treatment, into rivers, lakes and the sea. This still happens in developed countries, and is almost standard practice in the developing world, where many major rivers downstream from large cities are little more than open sewers.

As everyone knows, dirty water kills. And on our planet it does so on a massive scale. The major killer is common enough: diarrhoea. In developing countries some 4 million people are affected each year and many

of them, mostly children, die as a result of the dehydration that follows severe and untreated diarrhoea. Such disease is easily treated with salts dissolved in water. Even so, the World Health Organization estimates that unsafe water, sanitation and hygiene now cause the death of nearly 200 people an hour – that is, 200 people every hour of every day of every year.

But this is not the end of the story. As much as 80 percent of all illnesses in developing countries are water-related. Some, like diarrhoea, are transmitted through water by micro-organisms. Others, including malaria, which still kills more than 1 million people per year, are transmitted by larger animals (in this case the mosquito) that live on or in stagnant water. The list is long: dengue fever, guinea worm disease, trachoma, bilharzia, and river blindness are the best known. Of these, many are debilitating diseases that condemn sufferers to a lifetime of blindness or inaction. The economic loss to countries that can ill afford it is staggering.

While action can, and should, be taken to control and eliminate all these diseases, contaminated water can cause even more intractable problems. A relatively recent discovery is the high level of arsenic found in water supplies in many different parts of the world. The situation has become extremely serious over large areas in Bangladesh, ironically as a result of a programme to sink more and deeper boreholes to obtain better quality water. The new water has been found to be contaminated, in some places, heavily, by naturally occurring arsenic. This, it must be stressed, is not the result of industrial contamination but of natural infiltration. Whatever the cause, the resulting disease, called arsenicosis, builds up over time and can cause death. Fluorosis, caused by the build-up of fluoride, is a similarly insidious form of water poisoning that has become a major public health problem in many parts of Asia and north Africa.

Water has much to answer for. Those whose job it is to manage it have even more to answer for. Thus far they have done a bad job. However, it is not too late to improve the situation, as shown in Part 6.

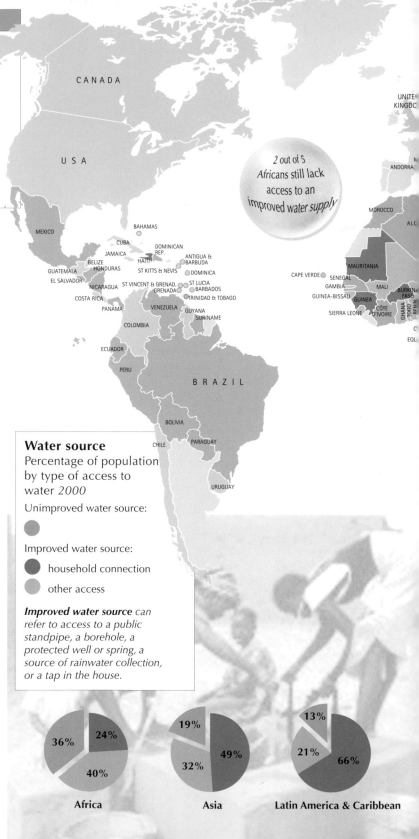

*2 out of 5
Africans still lack
access to an
improved water supply*

M ost people in the world do not have a cold water tap in their house. Instead, they have to fetch their water, often several times a day, in a pail or jar. Women may carry around 15 litres of water at a time – back-breaking work, which may take several hours each day.

Governments and aid agencies have made significant efforts to improve access to fresh water. But although the number of people served with some form of improved water supply increased from over 4 billion in 1990 to nearly 5 billion in 2000, the increase in population means that the number of people without access to an improved supply remained at over 1 billion. Most of these people live in Asia and Africa, with rural services lagging far behind those in urban areas.

Where water has to be carried into the house, people use it sparingly, and their hygiene and health suffer accordingly. In Swaziland, for example, people in households with taps use between 30 and 100 litres a day, while those who have to pay for water to be delivered use only 13 litres a day. People who have to carry water to their home, however, use only 5 litres a day — less than that used in one flush of a modern lavatory. Although 5 litres is sufficient for drinking, it is not nearly enough for washing bodies and clothes, cooking and cleaning dishes as well.

Water source

Percentage of population
by type of access to
water *2000*

Unimproved water source:

Improved water source:
- household connection
- other access

Improved water source can refer to access to a public standpipe, a borehole, a protected well or spring, a source of rainwater collection, or a tap in the house.

Africa
- 36%
- 24%
- 40%

Asia
- 19%
- 32%
- 49%

Latin America & Caribbean
- 13%
- 21%
- 66%

ACCESS TO WATER

Over 1 billion people are still without easy access to a reliable source of water.

145 hours per person per year could be saved in Africa by increasing piped water systems and sewer connections

Urban and rural facilities
Percentage of population
with access to an improved water source
2000

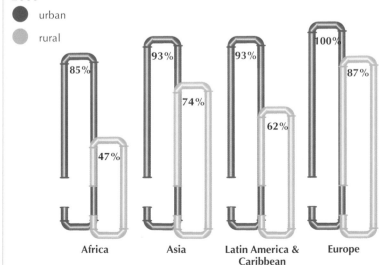

- ● urban
- ● rural

	Africa	Asia	Latin America & Caribbean	Europe
urban	85%	93%	93%	100%
rural	47%	74%	62%	87%

In Africa, maintaining a supply of water for the household is primarily the responsibility of women, and many spend up to five hours each day collecting it. Millions have to walk a long distance to a source of water, often accompanied by their small children. Girls old enough to attend school may have to rise before dawn to fetch water before setting off for the classroom, making them late for school and too tired to study properly. The source of water may be an open pool, used by animals, or a dangerously deep hand-dug well, into which they have to climb. Even when an improved water source – a public well or a tap – is installed, women may still have to walk some distance to fetch their daily water.

49

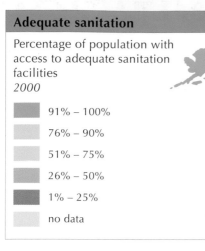

O ver a third of the world's population is still living with inadquate sanitation facilities. The safe disposal of human faeces is a key factor in the fight against many infectious diseases, but untreated sewage is a continuing health problem.

Many schools in developing countries have no toilets, and yet the provision of facilities that afford children some privacy is a major factor in encouraging more girls, in particular, to attend school. The education of girls, in turn, is key to a country's economic and social development.

Improving sanitation is not always straightforward. For example, in areas where water has to be collected and carried, the introduction of pour-flush latrines increases the workload of the community's women.

Only a small fraction of the wastewater collected through sewage systems in developing countries is treated and disposed of properly. Most is discharged untreated into rivers, lakes and oceans, thereby undermining the potential health benefits of the improved facilities.

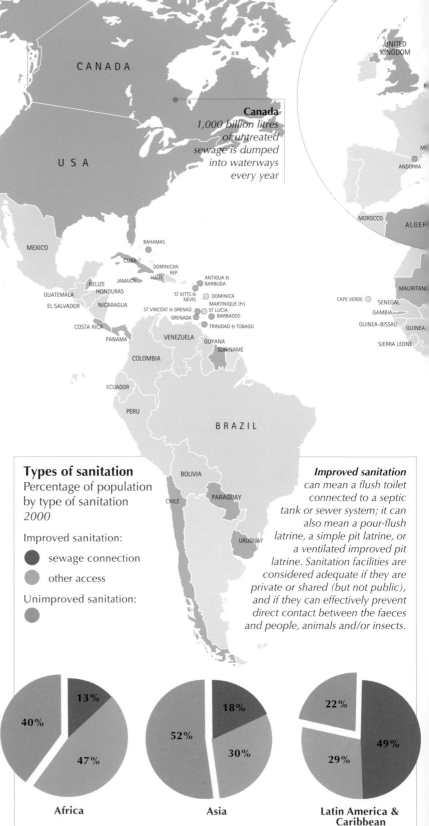

Canada
*1,000 billion litres
of untreated
sewage is dumped
into waterways
every year*

Types of sanitation
Percentage of population
by type of sanitation
2000

Improved sanitation:
- sewage connection
- other access

Unimproved sanitation:

Improved sanitation
*can mean a flush toilet
connected to a septic
tank or sewer system; it can
also mean a pour-flush
latrine, a simple pit latrine, or
a ventilated improved pit
latrine. Sanitation facilities are
considered adequate if they are
private or shared (but not public),
and if they can effectively prevent
direct contact between the faeces
and people, animals and/or insects.*

Africa — 13%, 47%, 40%

Asia — 18%, 30%, 52%

Latin America & Caribbean — 22%, 29%, 49%

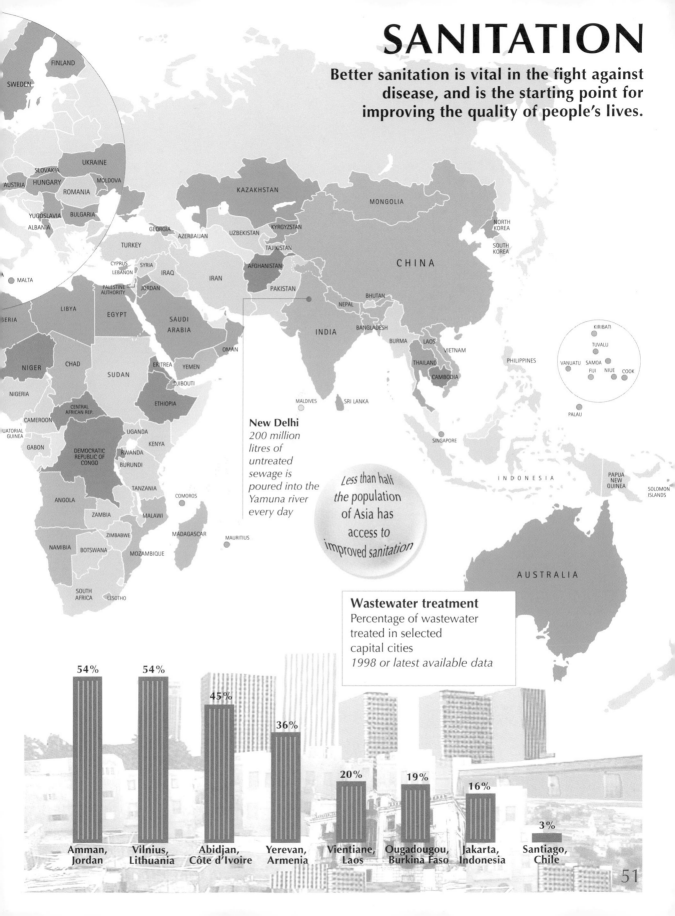

SANITATION

Better sanitation is vital in the fight against disease, and is the starting point for improving the quality of people's lives.

New Delhi
200 million litres of untreated sewage is poured into the Yamuna river every day

Less than half the population of Asia has access to improved sanitation

Wastewater treatment
Percentage of wastewater treated in selected capital cities
1998 or latest available data

54% — Amman, Jordan
54% — Vilnius, Lithuania
45% — Abidjan, Côte d'Ivoire
36% — Yerevan, Armenia
20% — Vientiane, Laos
19% — Ougadougou, Burkina Faso
16% — Jakarta, Indonesia
3% — Santiago, Chile

The bacteria and other agents that cause infectious diseases such as amoebic dysentery, cholera, typhoid and polio, are easily passed on in drinking water that is contaminated by human or animal faeces.

Diarrhoea is endemic in countries with poor water supplies. Children become infected and then re-infected in a deadly cycle of disease. Severe diarrhoea, if left untreated, can kill simply through dehydration. Chronic diarrhoea, to which malnourished children are especially vulnerable, can itself be the cause of malnutrition.

Some water-borne diseases do not kill immediately, but debilitate the sufferer, leaving them susceptible to further disease and unable to work to support themselves or their family. This is the case with schistosomiasis (bilharzia), which is caused by a parasite, and affects some 200 million people. It causes chronic illness and stunts a child's growth and development.

All water-borne diseases are exacerbated by, and are in turn the cause of, poverty. It is no surprise to find that the burden of water-borne disease falls almost exclusively on the developing world, where an estimated 82 million years of healthy life are lost each year.

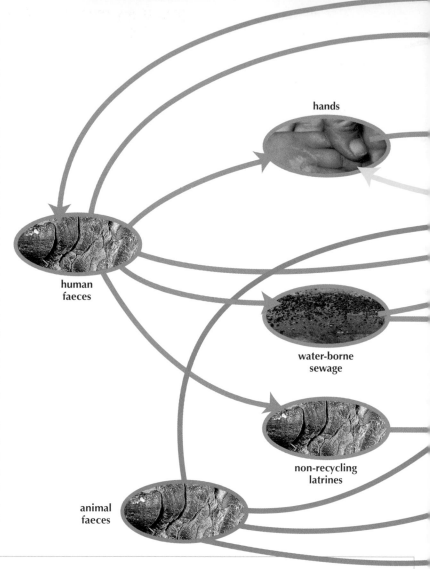

hands

human faeces

water-borne sewage

non-recycling latrines

animal faeces

Major water-borne diseases

Cholera
Cholera is an acute infection of the intestine that can cause severe vomiting and diarrhoea. Outbreaks occur sporadically throughout the world, especially in overcrowded communities with poor sanitation and unsafe drinking water. It affects malnourished people especially badly. Left untreated, severe cholera causes death through rapid dehydration in 50% of sufferers, but prompt treatment can reduce the death rate to less than 1%. In 2001, more than 184,000 cases of cholera were reported to the World Health Organization.

Typhoid
Typhoid is caused by a bacterial infection passed on in drinking water, or on food handled by an infected person. It can produce high fever and other symptoms. Around 17 million people are thought to be infected each year.

Guinea-Worm Disease
Guinea-worm disease is caused by a large roundworm, the larvae of which may be present in drinking water. The larvae develop inside people, and emerge from the legs and feet as a mature worm up to 1 metre long. As well as unpleasant ulcers the disease can cause debilitating symptoms that prevent people from working and producing food.

Diarrhoea
Diarrhoea is usually a symptom of a bacterial, viral or parasitic infection, most commonly spread by contaminated water. Severe diarrhoea can cause death from dehydration. Chronic diarrhoea is both caused by, and the cause of, malnutrition.

Polio
Polio is caused by a virus passed on in drinking water, food or by human contact. Irreversible paralysis occurs in 1 in 200 cases, and can lead to death when breathing is affected. Vaccination has successfully reduced the number of reported cases by 99% since the 1950s, but the virus continues to be a problem in parts of Africa and South Asia.

DIRTY WATER KILLS

Dirty water causes 1.7 million deaths each year. This is the equivalent of 10 jumbo jets crashing every day; 90% of the passengers are children.

flies

soil

food

people

surface water

drinking water

groundwater

The cycle of disease

Poor sanitation and lack of sewage treatment can lead to soil, surface water and groundwater becoming contaminated by disease-causing agents present in human and animal faeces. These are then passed on through drinking water, through the water that is used to grow food, and through human contact with the food itself. People also become infected through direct contact with water when they are bathing, or simply collecting water.

80% of illnesses in developing countries are water-related

Trachoma
Number of useful years lost because of blindness caused by trachoma
2001

Some diseases flourish because water for washing is in short supply. Trachoma, which causes blindness, is transmitted by physical contact between people, and is often passed on from child to mother. It affects nearly three times as many women as men.

People who are blind find it hard to work and support themselves. The effect of trachoma can be measured in terms of the years of potentially active and productive life that is lost.

total
3.95 million years

Western Pacific
1.6 million

Southeast Asia
248,000 years
Eastern Mediterranean
602,000 years

Africa
1.5 million years

Deaths from dirty water
Number of deaths attributable to unsafe water, sanitation and hygiene
2000
affected countries

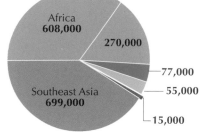

Africa
608,000

270,000

77,000

55,000

Southeast Asia
699,000

15,000

Total deaths from dirty water: 1.724 million

EUROPE

WESTERN PACIFIC

EASTERN MEDITERRANEAN

AMERICAS

AFRICA

SOUTHEAST ASIA

53

More than a million people die each year from malaria. The disease is spread by mosquitoes, which lay their eggs in standing water. Many other vector-borne diseases, such as the disabling elephantitis, which causes the limbs to swell, are endemic in tropical regions, and some, such as West Nile Virus, are spreading northwards to affect the industrial world.

The provision of closed water sources such as pipes and taps helps reduce the incidence of these diseases, but if the water supply is only intermittent, people may resort to storing water, creating ideal breeding grounds for insects.

Water collection and storage is essential in arid regions, as is the channelling of water to fields to irrigate crops and increase food production. But the very technology intended to improve people's quality of life brings with it further problems. Dams, both large and small, create areas of still water behind them, and these have been shown to encourage certain diseases.

In Ethiopia micro-dams, considered to be more environmentally sustainable than large-scale operations, have led to a seven-fold increase in the incidence of malaria. In northern Senegal, an area where schistosomiasis was previously unknown, the building of the Diama dam in 1986 led to almost the entire population being infected by 1994.

Malaria affects 300 million people worldwide. Poverty, inadequate water resources and underfunded health services make it difficult to eradicate. The disease, in turn, places a huge financial burden on the poor. It is estimated that each bout of malaria costs 10 working days, and that malaria accounts for up to 30 percent of hospital admissions in tropical Africa.

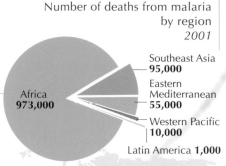

Malaria
Number of deaths from malaria by region
2001

Africa **973,000**

Southeast Asia **95,000**

Eastern Mediterranean **55,000**

Western Pacific **10,000**

Latin America **1,000**

Total deaths from malaria: 1.134 million

87% of deaths from malaria occur in Africa

60% of deaths from DHF occur in Southeast Asia

Dengue is a mosquito-borne infection found in tropical and sub-tropical regions, predominantly in urban areas. In 1970, it was known only in nine countries, but it is now endemic in more than 100. It is thought to affect 50 million people each year. In 2001, Brazil reported more than 390,000 cases.

Dengue haemorrhagic fever (DHF) is a potentially lethal complication of dengue infection. If left untreated the fatality rate can exceed 20 percent.

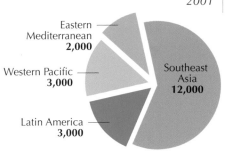

Dengue
Number of deaths from dengue haemorrhagic fever (DHF) by region
2001

Eastern Mediterranean **2,000**

Western Pacific **3,000**

Latin America **3,000**

Southeast Asia **12,000**

Total deaths from DHF: 20,000

Blighted lives
Number of useful years lost because of disabling illnesses
2001

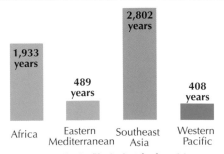

Lymphatic filariasis (elephantitis)

Africa	Eastern Mediterranean	Southeast Asia	Western Pacific
1,933 years	489 years	2,802 years	408 years

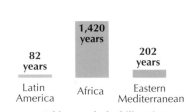

Schistosomiasis (bilharzia)

Latin America	Africa	Eastern Mediterranean
82 years	1,420 years	202 years

HARBOURING DISEASE

Water is still a breeding ground for some of the world's deadliest killers.

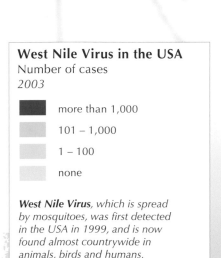

West Nile Virus in the USA
Number of cases
2003

- **more than 1,000**
- 101 – 1,000
- 1 – 100
- none

West Nile Virus, *which is spread by mosquitoes, was first detected in the USA in 1999, and is now found almost countrywide in animals, birds and humans.*

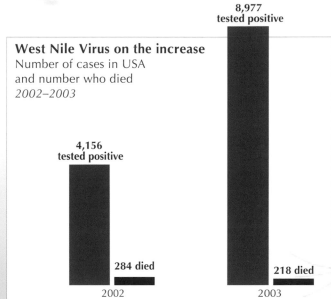

West Nile Virus on the increase
Number of cases in USA
and number who died
2002–2003

8,977
tested positive

4,156
tested positive

284 died

218 died

2002

2003

WASHINGTON

MONTANA

NORTH DAKOTA

MINNESOTA

MAINE

OREGON

IDAHO

WYOMING

SOUTH DAKOTA

WISCONSIN

VERMONT
NEW HAMPSHIRE

MASS.

NEVADA

UTAH

NEBRASKA

IOWA

MICHIGAN

NEW YORK

CONN. RHODE ISLAND

PENNSYLVANIA

NEW JERSEY

CALIFORNIA

COLORADO

KANSAS

ILLINOIS

INDIANA

OHIO

WASHINGTON D.C.

WEST VIRGINIA

VIRGINIA

DELAWARE
MARYLAND

MISSOURI

KENTUCKY

NORTH CAROLINA

ARIZONA

NEW MEXICO

OKLAHOMA

ARKANSAS

TENNESSEE

SOUTH CAROLINA

TEXAS

MISSISSIPPI

ALABAMA

GEORGIA

LOUISIANA

FLORIDA

Water that appears to be clean may be toxic. Even water from an 'improved water supply' – a well or a tap – may contain deadly poisons, some introduced by human activity, but others simply present in the ground.

Nitrates are widely used as a fertilizer but can cause 'blue baby syndrome', when infants convert nitrate into nitrite, which stops their blood from transporting oxygen and can lead to suffocation and death. Since 1950, around 3,000 deaths have been reported worldwide, but in many countries the syndrome goes unrecognized or unrecorded. In the USA up to 27,000 babies may be at risk, and several lawsuits have been taken out against water companies.

Chemicals such as organochlorines and phyto-oestrogens are present in industrial, agricultural and municipal waste. Their effect on the human hormonal system is still being assessed, but they have been linked to falling sperm rates and other abnormalities.

The pesticide DDT was once widely used. Now banned in many countries, its residues are still widespread. Although known to affect the nervous system in high doses, DDT is still used against mosquitoes in the battle to control malaria.

Arsenicosis, or arsenic poisoning, can take 5 to 20 years to develop. It leads to skin problems, cancers, diseases of the blood vessels of the legs and feet, and possibly causes diabetes, high blood pressure and reproductive disorders.

Many of its acute symptoms can be cured by giving the sufferer a supply of healthy water.

Arsenic in the water

arsenic-contaminated water is a problem in some areas

Arsenic in the USA
Around 13 million people drink water tainted with high levels of arsenic in parts of California, Minnesota, Nevada, Oregon and Texas

Fluoride is added to the water in some countries in microscopic proportions in order to improve dental health, but in excessive amounts it can cause disease and even death.

It occurs naturally in the water in certain areas, and can lead to 'skeletal fluorosis', in which the bone structure is changed and ligaments calcify.

There is no treatment for the disease. Removing fluoride from the water is difficult, expensive and produces a residue with high concentrations of fluoride, which then has to be disposed of safely.

Fluoride

fluorosis is an endemic public health problem

Fluorosis
is thought to affect around 70 million people in northern China, and 30 million people in northwest India

INSIDIOUS CONTAMINATION

Microscopic amounts of chemicals in drinking water can build up in the body and have a devastating effect on health.

Lead levels in some of Asia's rivers are 20 times higher than in rivers in OECD countries

Up to half the population in Bangladesh is at risk from high levels of arsenic in drinking water. In the 1970s millions of deep wells were dug, intended to provide Bangladeshis with clean, safe drinking water. But arsenic in the ground is finding its way into the wells – a situation only recognized in the mid-1990s.
In some Bangladeshi villages, where a new well is producing uncontaminated water, villagers are asked to conserve the safe water by using it only for drinking and cooking.

Lead

Although lead is known to impair brain function, old lead piping still carries water into houses in the industrialized world

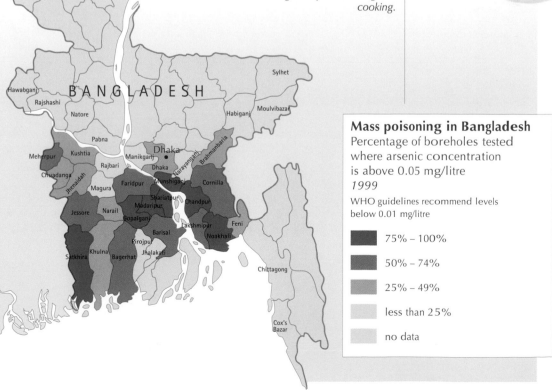

Mass poisoning in Bangladesh
Percentage of boreholes tested where arsenic concentration is above 0.05 mg/litre
1999

WHO guidelines recommend levels below 0.01 mg/litre

- 75% – 100%
- 50% – 74%
- 25% – 49%
- less than 25%
- no data

PART 4

RE-SHAPING THE NATURAL WORLD

FEW OF THE WORLD'S RIVERS exist in their natural state. Nearly all have been dammed, some have been diverted, and others have been drained of their natural resource before it ever reaches the sea. Currently, humans make use of some 54 percent of all available fresh water from rivers, lakes, streams and shallow aquifers. More than 40,000 large dams bisect the world's waterways, and more than 500,000 kilometres of river have been dredged for shipping.

Plans to re-design water systems on a grand scale are being mooted in China and in Spain. In India, a controversial scheme to link the river system of the entire subcontinent, thereby bringing water from the Himalayas to the arid south, is causing disputes between states and may well founder. Equally ambitious plans have never come to fruition. In North America in the 1960s, plans were made to divert water from Canadian and Alaskan rivers to central Canada, the south-west of the USA and northern Mexico. Most of the water would eventually have drained away into the Californian desert, a fate which Canadians may have found a less than useful diversion for one of their major natural resources. In Russia, a plan to reverse the flow of three north-flowing rivers to send water to central Asia would have eclipsed even the disastrous decline of the Aral Sea. The project was cancelled in 1986.

Unhappily, plans to drain and build on the world's wetlands have not been cancelled in the same way. Wetlands have often been seen as wastelands, serving no useful purpose. About half of them have disappeared in the past 100 years. In some areas of Europe, such as Germany and France, 80 percent of all wetlands have now been destroyed. As a result, 20 percent of all freshwater fish are threatened or extinct, and many amphibians are in decline.

Wisdom is finally prevailing, and a good thing too, because wetlands do far more than support wild bird populations. They also play a crucial role in regulating the world's water systems. They are the world's natural water filters, and can even be used, more or less without interference, to process sewage.

Deep underground, though, the plunder continues. Fossil waters, tens of thousands of years old, are being pillaged to support wasteful agricultural expansion and the vociferous demands of thirsty city dwellers. In Libya, for example, 90 percent of the country's water is extracted from underground aquifers. Many of the world's cities depend primarily on the water beneath them for safe supplies of drinking water. Not surprisingly, this 'safe' water is becoming polluted and its volume is shrinking. As it does so, subsidence leaves its tell-tale mark on the falling levels of city streets.

The situation is particularly serious in India. Overpumping of groundwater in Gujarat has resulted in salt water invading the aquifers and contaminating drinking water supplies. Further south in Tamil Nadu, water tables have dropped by as much as 30 meters since the 1970s.

Such calamities engender desperate measures, but few are yet available. Desalination has helped a number of countries, mostly in the Middle East, but the quantity of sea water turned into fresh water is still minute. The world's water tankers plod a weary path from water-rich countries to such places as the Bahamas, Antigua, Mallorca, South Korea, Taiwan and the Pacific Islands — and while their deliveries improve the lives of isolated island communities, they make little impact on the world water problem. Transporting water in huge floating bags and towing icebergs from the Antarctic to the Sahara are still projects either for the future or for science fiction.

Behind all this lies the daunting prospect of global climate change. As we burn more fossil fuels, the greenhouse effect warms the planet and increases evaporation rates. Since what goes up must come down, rainfall increases and extreme climatic events become more common. One result is that floods become more frequent and more serious, ruining the lives of millions of people every year. Perversely, global warming may also increase drought in arid areas, particularly near the equator. Thus do the four horses of the Apocalypse – war, famine, pestilence and death – gallop even faster in times of water shortage and climate change.

More than 40,000 dams fragment the world's rivers

The damming of rivers for hydropower, water for irrigation schemes, flood control and improved navigation, have left some rivers, such as the Mississippi, little more than a series of connected lakes.

This process of 'fragmentation' can have a serious effect on the ecology of a river, and on the people who live downstream and depend on its waters, but even more grandiose schemes are now being planned to divert rivers from one region to another.

Around the world, diversion schemes are being proposed to alleviate water shortages, inevitably creating conflict, as one group of people sees its vital resource being channelled away for the benefit of another (*see page 78*).

Nepal: Melamchi Project

— river
— tunnel
— distribution pipes
■ water treatment plant
□ reservoir
▒ road
⋯ road under construction
▓ national park

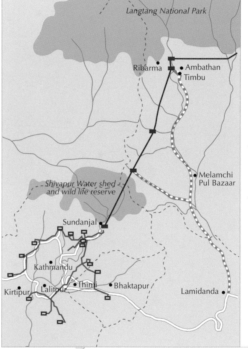

Water for Kathmandu

Most of the 1.5 million residents of Kathmandu Valley receive piped water for only two hours every other day. Some get none at all. People have to use expensive water brought by tankers, or water from unsafe wells. The Melamchi project involves diverting the river from an adjacent valley through a tunnel 26 kilometres long, treating it and distributing it to the three towns in the valley. Much of the estimated cost of US$464 million is being met by the Asian Development Bank, the World Bank and other foreign donors. The projected completion date is 2006. The residents of the Melamchi valley, who will undoubtedly be affected by the diversion of some of their river, are being provided with improved health and education facilities, and electrification.

Spain: Proposed diversion of the Ebro

↙ direction of diversion and quantity of water diverted
million m³ per year

Current availability of water:

▓ excess
▒ equilibrium
▒ deficit
▓ major deficit

Cataluña
190 million m³ per year

Júcar
315 million m³ per year

Segura
450 million m³ per year

Almería
95 million m³ per year

Water for farmers

The Spanish government is planning to create 828 kilometres of new waterways and a number of dams in order to divert some of the waters of the Ebro to alleviate water shortages in the south of the country. Critics point out that a better plan would be to encourage farmers to use less water by removing water subsidies. They question whether it is sensible to grow water-needy crops in water-starved regions. Environmentalists are concerned about the effect of the diversion on the Ebro delta.

DIVERTING THE FLOW

Few of the world's major rivers flow freely. Most have been harnessed to provide energy and irrigation. Some are being diverted hundreds of kilometres from their natural course, with devastating environmental and human costs.

Water for coal

A shortage of water in Shanxi province has cost China US$600 million a year in lost coal production, and restricted the citizens of Taiyuan to only 25 litres of water a day. Diversion of water from the Yellow River to Taiyuan through 300 km of tunnels to an elevation of 636 metres has helped alleviate the problem, but will exacerbate shortages already experienced downstream.

Water for the north

Industry makes a greater profit out of water than agriculture does, which is why China is diverting water to service the needs of its urban and industrial centres.
The proposed solution to the now regular failure of the Huang He (Yellow River) to reach the sea for a part of every year, is a massive scheme involving three south-to-north canals that will link China's four large rivers into a single network. The canals, when completed in 2050, will be able to carry up to 44.8 billion cubic metres of water a year – similar to the flow of the Rhine in western Europe – supplementing the meagre flow of the Yellow River with the more plentiful waters of the Yangtze. The eastern route will be finished in 2007. It uses mainly existing river channels, with the addition of 30 pumping stations. The middle route, which involves raising the height of the Danjiangkou Dam by 23 metres and building 1,200 kilometres of canals, will not be completed until about 2030. The western route – a major engineering challenge – will involve the construction of dams 200 metres high, and tunnels more than 100 kilometres long, at altitudes of 3,000 to 5,000 metres. Critics point to the estimated 250,000 people who will have to be relocated, and to the fact that the project will meet only a small proportion of the north's water deficit. Clearly, more radical water-saving solutions are needed.

China: Water for the north

— river
----- provincial boundary

South-to-North water diversion:
◄— eastern route
◄— middle route
◄— western routes

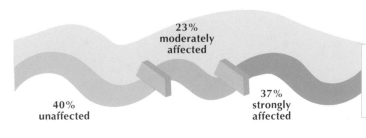

40% unaffected

23% moderately affected

37% strongly affected

River fragmentation
Percentage of the world's large river basins affected by dams and diversions
2000

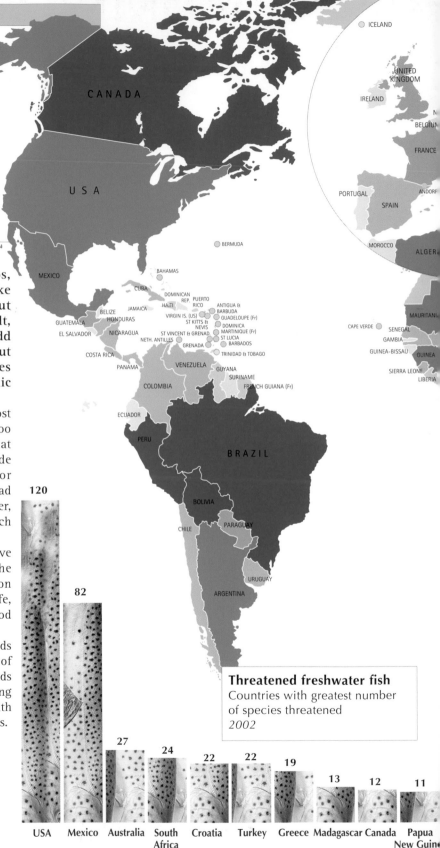

Wetlands – bogs, swamps, and marshes – act like sponges. They filter out impurities in the water, absorb silt, regulate the flow of water and add moisture to the atmosphere. Without them, rivers flow too fast, lakes become overburdened with organic matter, and coastlines are eroded.

Half of the world's wetlands were lost in the last century, when they were too often perceived as waste ground that only required draining to provide much-needed extra space for development. A fear of malaria, spread by mosquitoes breeding in still water, provided a further rationale for much wetland clearance.

The destruction of wetlands can have a devastating impact on the environment, including the extinction of freshwater fish and other wildlife, and the consequent loss of livelihood for local people.

In 1971, the Convention on Wetlands was signed at Ramsar in Iran. A list of 1,388 important international wetlands has been drawn up over the intervening years by the 138 signatory states, with the aim of conserving these vital areas.

Threatened freshwater fish
Countries with greatest number
of species threatened
2002

120	82	27	24	22	22	19	13	12	11
USA	Mexico	Australia	South Africa	Croatia	Turkey	Greece	Madagascar	Canada	Papua New Guinea

DRAINING WETLANDS

Wetlands help keep the world's freshwater resources healthy, but all too often they are seen as wastelands that can be drained and developed to house growing populations.

Oasis of Azraq
1977: ancient oasis made Ramsar site
1980: Jordan began to pump out water
1993: oasis reduced to dust

Garden of Eden

The destruction of the marshlands of southern Iraq

● extent of marshland area in 1973

● existing marshland in 2002

The marshlands of Southern Iraq – dubbed the 'Garden of Eden' – have been almost completely destroyed. More than 30 large dams fragment the Tigris and Euphrates, reducing the amount of water available downstream. But it was the extensive drainage schemes ordered by Saddam Hussein in the early 1990s that proved the final straw. The wetlands – once a richly diverse habitat – have largely been turned to desert, and their inhabitants – 500,000 Marsh Arabs, or Madan – displaced. After the fall of Saddam in 2003, pledges were made to restore the marshes, but this is likely to be a long-term project.

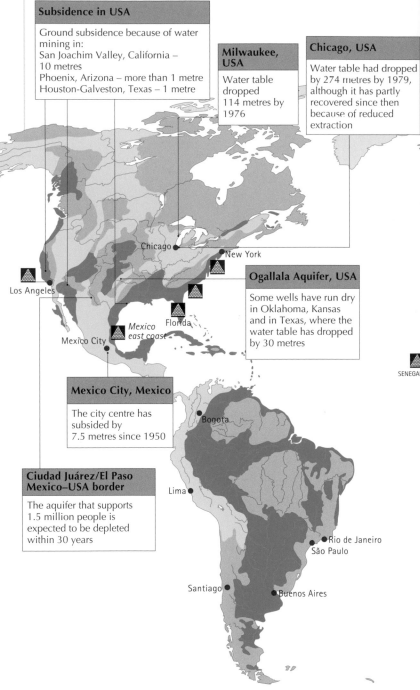

Subsidence in USA

Ground subsidence because of water mining in:
San Joachim Valley, California – 10 metres
Phoenix, Arizona – more than 1 metre
Houston-Galveston, Texas – 1 metre

Milwaukee, USA

Water table dropped 114 metres by 1976

Chicago, USA

Water table had dropped by 274 metres by 1979, although it has partly recovered since then because of reduced extraction

Ogallala Aquifer, USA

Some wells have run dry in Oklahoma, Kansas and in Texas, where the water table has dropped by 30 metres

Mexico City, Mexico

The city centre has subsided by 7.5 metres since 1950

Ciudad Juárez/El Paso Mexico–USA border

The aquifer that supports 1.5 million people is expected to be depleted within 30 years

Chicago • New York
Los Angeles
Mexico east coast
Florida
Mexico City
SENEGAL
Bogota
Lima
Rio de Janeiro
São Paulo
Santiago
Buenos Aires

We are withdrawing water from underground aquifers at a faster rate than it can be replenished. In some countries, this water is used to irrigate crops to feed hungry populations. In others, it is being used to fill swimming pools, water golf courses and service holiday resorts.

Underground aquifers hold almost all the fresh water that is not in the form of ice. The water in some aquifers is millennia old and lies beneath what are now some of the driest regions on Earth. Although people have drawn water from springs and wells since the earliest civilizations, in the past 50 years multiplying populations have needed more food and water, and the rate of withdrawal has increased dramatically.

In some coastal areas, so much fresh water has been withdrawn from aquifers that saltwater has started to intrude, turning well water brackish and unusable. In places such as Mexico City (*see page 67*), the emptying of aquifers has caused serious subsidence.

Some aquifers are refilled by rainwater soaking into the ground, but this process may take hundreds, or even thousands, of years, making such water effectively a non-renewable resource.

GROUNDWATER MINING

The world's aquifers are being mined for their valuable treasure. Although immense, they are not bottomless, and in many areas water levels are sinking fast.

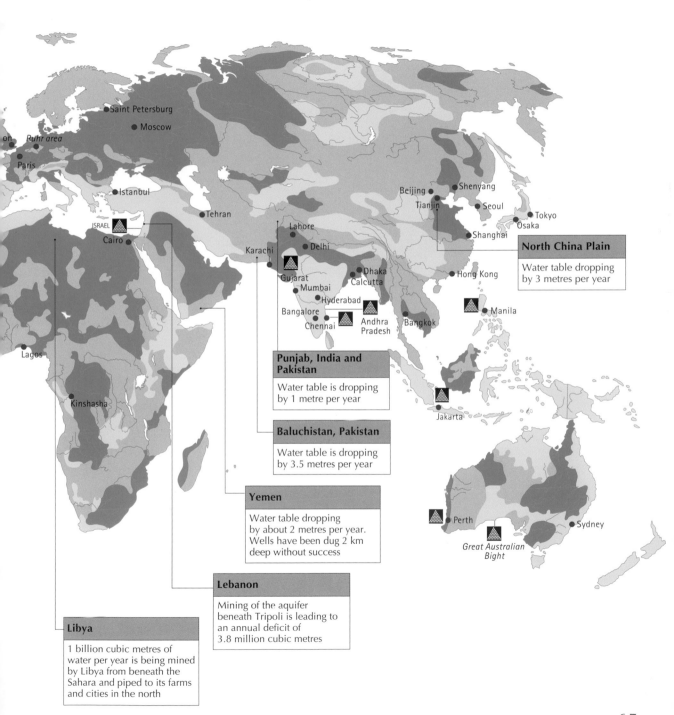

North China Plain

Water table dropping by 3 metres per year

Punjab, India and Pakistan

Water table is dropping by 1 metre per year

Baluchistan, Pakistan

Water table is dropping by 3.5 metres per year

Yemen

Water table dropping by about 2 metres per year. Wells have been dug 2 km deep without success

Lebanon

Mining of the aquifer beneath Tripoli is leading to an annual deficit of 3.8 million cubic metres

Libya

1 billion cubic metres of water per year is being mined by Libya from beneath the Sahara and piped to its farms and cities in the north

U rban areas are among the world's most life-threatening environments. The high concentration of people, coupled with inadequate provision of water and sanitation, provides a perfect breeding ground for infectious diseases. And how much worse would conditions be if water supplies were to run out?

Water is fundamental to health, and the installation of a plentiful, reliable supply of water in an area of a city is one of the most effective ways of improving the health and well-being of a large group of people. This shows up most clearly in child mortality rates. In many middle-income countries, deaths of children in urban areas, where water supplies may be unreliable, are as high as 50 to 100 per 1,000 live births, as against an average of 39 for middle-income countries as a whole.

The problem is not just how to get water to the people, but, for many cities, where to get the water from. Many of the world's largest cities – Los Angeles, Mexico City, Cairo, Calcutta, Beijing – are located in water-stressed areas, and they are struggling to meet the water needs of their growing populations.

Urban water safety
Percentage of drinking-water tested and found to violate national standards
2000

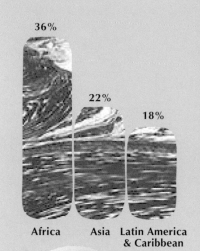

36% Africa
22% Asia
18% Latin America & Caribbean

Wasted water
Percentage of water lost through leaking pipes in large cities
2000

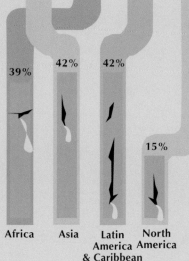

39% Africa
42% Asia
42% Latin America & Caribbean
15% North America

No sanitation
In Africa and Asia 60 percent of city dwellers live without adequate sanitation

Mexico City growth
Increasing number of people
1940–2000

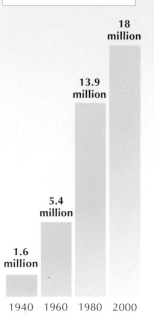

18 million
13.9 million
5.4 million
1.6 million

1940 1960 1980 2000

Urbanization
By 2015 nearly half the people in the developing world are expected to live in cities

City living
Number living in urban areas
1975–2000, 2015 projected
millions

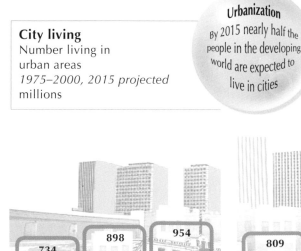

734 898 954
more developed regions
1975 2000 2015

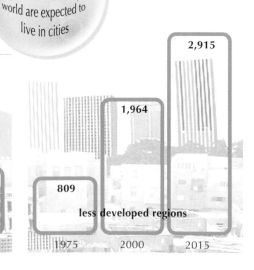

809 1,964 2,915
less developed regions
1975 2000 2015

EXPANDING CITIES

The rapid growth of cities is putting an increasing strain on already over-stretched water resources.

6 million
1945

26 million
1997

City growth in USA
Urban land area
1945–1997
hectares

Urban areas interfere with the water cycle. Rain falling on paved streets is channelled into gutters and sewers. Much of this is discharged into rivers or directly into the sea instead of being allowed to sink through the ground and recharge aquifers. In this way, the amount of fresh water is being depleted. Inland, the surge of water produced by heavy rainfall can flood rivers, and carries with it chemicals, organic matter and silt.

Mexico City's water sources

26% imported water

72% Mexico basin aquifer

2% Mexico basin other

Imported water is pumped up a height of 1,200 metres, over a distance of about 180 km from the Cutzamala and Lerma river basins

Mexico City's wastewater treatment

25% treated

75% untreated

Mexico City is a good example of a city trying to provide adequate water and sanitation for its rapidly expanding population, largely from a non-renewable resource.

Nearly three-quarters of the city's water comes from the aquifer over which it is built. At current rates of use – 15 million cubic metres a day – there is enough water left for 150 to 200 years, but demand is increasing, and as the level drops, the water will become harder to extract and its quality will decline. Once it is gone, there will be very little water to support a population of over 18 million people.

For over a century, the city has been subsiding as a result of water being pumped from the aquifer. The centre of Mexico has dropped by 7.5 metres, and is now lower than Lake Texcoco. A drainage network is in place to reduce the risk of flooding, but pipes are liable to be cracked by the subsiding ground.

The citizens of Mexico City suffer from the infectious diseases common in developing countries, passed on in contaminated water. The city's water courses have been polluted with toxic chemicals, fertilizers and human waste. Only 25% of wastewater receives any treatment, and sewage leaking from pipes cracked by subsiding ground has contaminated the aquifer – the city's main water supply.

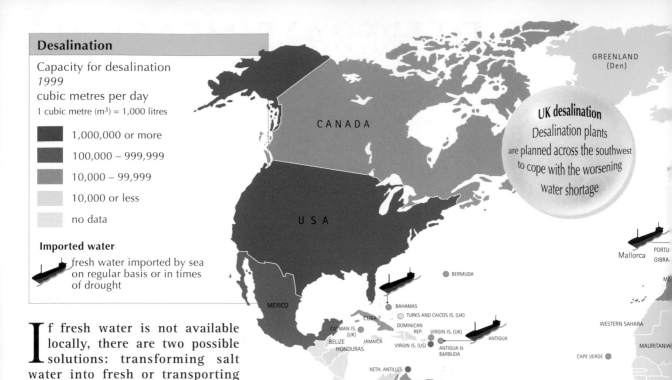

Desalination

Capacity for desalination
1999
cubic metres per day
1 cubic metre (m³) = 1,000 litres

- 1,000,000 or more
- 100,000 – 999,999
- 10,000 – 99,999
- 10,000 or less
- no data

Imported water
fresh water imported by sea
on regular basis or in times
of drought

UK desalination
Desalination plants
are planned across the southwest
to cope with the worsening
water shortage

If fresh water is not available locally, there are two possible solutions: transforming salt water into fresh or transporting fresh water from wetter places.

Desalination (the conversion of salt water or wastewater into fresh) only accounts for about 1 percent of world water consumption, mainly because it requires expensive technology and a massive amount of energy. But in oil-rich countries it is a viable option, and Kuwait and Bahrain rely on it. Elsewhere, the technology is sometimes used to supply areas where water is in short supply, but it is beyond the means of many poorer economies.

Many water-stressed communities have to rely on motorized tankers to deliver fresh water overland. This is a risky business because the supplies can easily become contaminated – if they arrive at all.

Small island communities often need water to be brought by sea in tankers or barges. In Greece, floating polyurethane bags filled with up to 2 million litres of drinking water are towed out to some islands. Early in 2004, Israel and Turkey were near to finalizing a deal whereby Turkey would sell to Israel some 50 million cubic metres of water each year. Special tankers would be built by the Israelis to transport the water from Turkey's Manavgat river.

DESPERATE MEASURES

Transforming salt water into fresh water or transporting fresh water to where it is needed, are sometimes the only options in water-stressed lands.

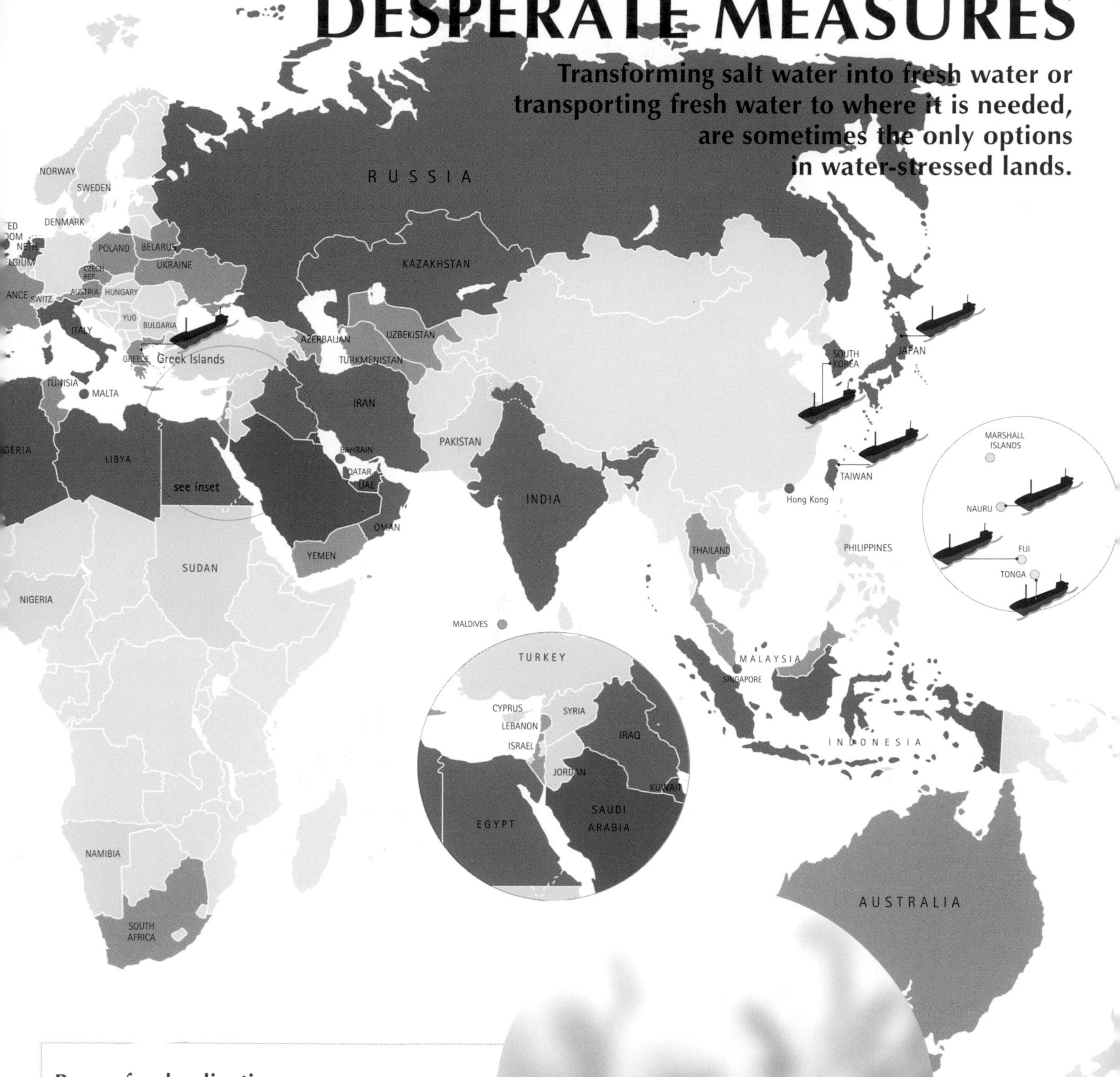

NORWAY
SWEDEN
DENMARK
ED
OOM
NETH
LGIUM
POLAND
BELARUS
CZECH
REP
UKRAINE
ANCE
AUSTRIA
SWITZ
HUNGARY
ITALY
YUG
BULGARIA
GREECE Greek Islands
TUNISIA
MALTA
GERIA
LIBYA
see inset
YEMEN
SUDAN
NIGERIA
NAMIBIA
SOUTH
AFRICA

RUSSIA

KAZAKHSTAN

AZERBAIJAN
UZBEKISTAN
TURKMENISTAN
IRAN
PAKISTAN
BAHRAIN
QATAR
UAE
OMAN

INDIA

SOUTH
KOREA
JAPAN
TAIWAN
Hong Kong
THAILAND
PHILIPPINES

MALDIVES

MARSHALL
ISLANDS
NAURU
FIJI
TONGA

TURKEY
CYPRUS
SYRIA
LEBANON
ISRAEL
IRAQ
JORDAN
KUWAIT
EGYPT
SAUDI
ARABIA

MALAYSIA
SINGAPORE

INDONESIA

AUSTRALIA

Power for desalination

Amount of power needed to turn saltwater into 1,000 litres of fresh water
2003
kilowatts per hour
1 kWh = power needed to run 1 bar on an electric fire for 1 hour

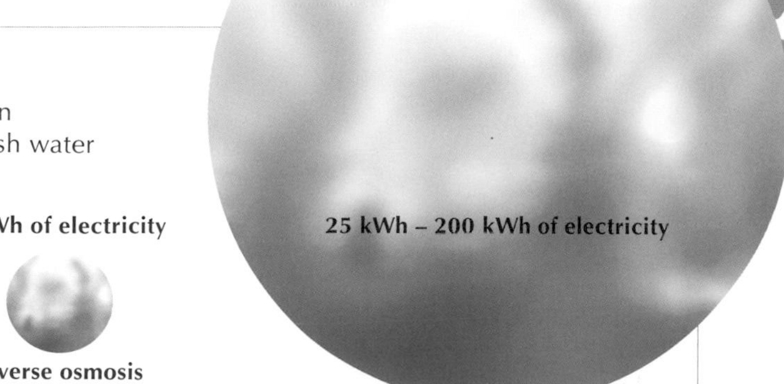

6 kWh of electricity

25 kWh – 200 kWh of electricity

reverse osmosis
involves pumps driven by electricity or diesel forcing water through a membrane to separate fresh water from salty water

distillation
involves heating water to create steam, which distills as fresh water

69

Rising floods
Number of worldwide floods
1992–2001

57 · 82 · 80 · 88 · 69 · 77 · 90 · 112 · 152

1992 1993 1994 1995 1996 1997 1998 1999 2000 2

venezuela 1998
Mudslides on
deforested slopes
killed 30,000 *people*

Death and disaster
Numbers killed
by floods and
financial cost
by continent
1992–2001

EUROPE
1,362 killed
costs: US$32,000m

ASIA
50,034 killed
costs: US$105,000m

AMERICAS
35,848 killed
costs: US$31,000m

AFRICA
9,243 killed
costs: US$892m

OCEANIA
20 killed
costs: US$792m

F loods affect millions of people each year. Often considered 'natural' disasters, many are made far worse by deforestation, draining of wetlands, and attempts to constrain river flow.

Climate change is leading to more extreme weather – heavier monsoon rains, and more powerful and more frequent cyclones and hurricanes. Rain falling on deforested slopes washes away the soil that would previously have soaked it up. This increases the amount of water running into rivers, and the amount of sediment. The Yangtze floods of 1998, which killed more than 4,000 people and cost in excess of US$36 billion, were made worse by logging above the river's upper reaches. The Chinese government has tried to restrict logging in the region and has started a major replanting scheme, but illegal logging is continuing.

Although improved flood warning systems are reducing the number of people killed, because land around rivers is increasingly drained and developed, millions remain in danger.

Nearly half of the Mississippi flows through artificial channels, and nearly 7 million hectares of wetlands – the river's natural sponge – have been drained for development. In 1993 the river breached over 10,000 kilometres of levees, and spread across 41,000 square kilometres, reclaiming its lost floodplain.

FLOODS

Floods kill thousands every year and ruin the lives of millions more. They are becoming more frequent.

Nearly 292 million people were affected by floods in 1998

Floods are almost an annual event in Bangladesh. The people living on the floodplain have adapted to cope with rising waters, which deposit fertile silt on their land, and provide areas for fish to breed. But the floods of 1998 were record-breaking, both in the amount of land affected, and the duration – more than two months. 1,300 people died, 31 million were left temporarily homeless, 16,000 km of roads were ripped up, and sea defences were destroyed.

The causes of the flooding are complex, but the human factors include deforestation in the Himalayas, and run-off in heavily developed areas downstream, both of which increased the flow of the Ganges and Brahmaputra rivers.

Ironically, riverbank defences, built to prevent flooding, lengthened the period of the disaster. Floodwater that had spilled over them and inundated fields, had no way of draining back into the river as the water level dropped.

Flooding in Bangladesh
October 1998

- area affected
- area severely affected

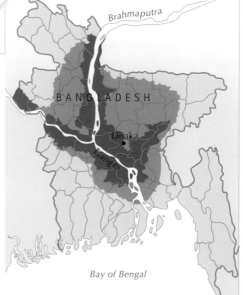

Bay of Bengal

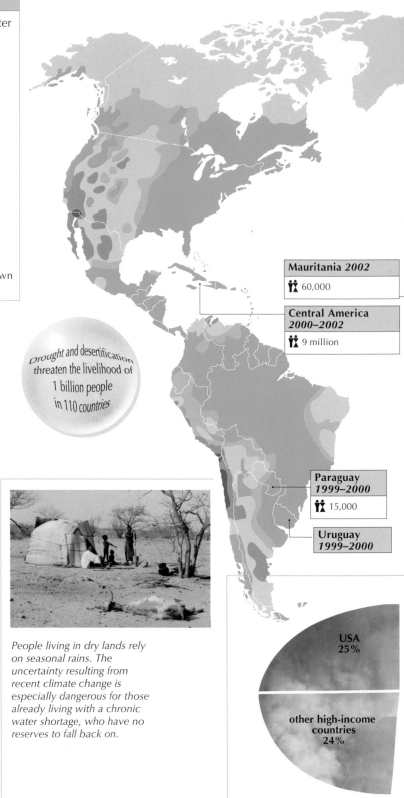

Aridity zones

Land classified according to amount of water lost to atmosphere as a proportion of total rainfall

- hyperarid
- arid
- semi-arid
- dry sub-humid
- moist sub-humid and humid
- cold

Life-threatening droughts

selected incidents reported by UN agencies 2000–03

👪 number of people affected, where known

Drought and desertification threaten the livelihood of 1 billion people in 110 countries

Mauritania 2002
👪 60,000

Central America 2000–2002
👪 9 million

Paraguay 1999–2000
👪 15,000

Uruguay 1999–2000

USA 25%

other high-income countries 24%

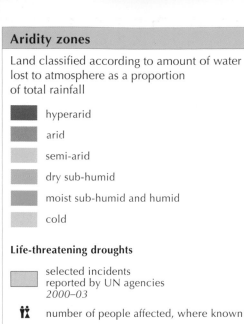

Droughts do not happen overnight; they may take several years to develop. A drought in one region might be three weeks without rain; in another it might be no rain for a year.

The billion or more people in the world's driest terrains, have largely adapted their way of life to long periods without rain. But they do rely on seasonal rains to water their crops; if these fail they face death from starvation and dehydration.

In the industrialized world, even plentiful water resources may be strained by the demands of agriculture, industry, and of millions of people crammed together in cities. Lower rainfall than usual is experienced as a drought because it affects daily life, even if nobody dies as a result.

Drought in arid areas, combined with misuse of the land, such as deforestation or over-grazing, exacerbates the process of desertification. Dried-out topsoil is simply blown away, and the land is permanently degraded.

Climate change, linked to the emission of greenhouse gases such as carbon dioxide (CO_2), is disrupting weather patterns. More rain is falling in some places, and less in others.

People living in dry lands rely on seasonal rains. The uncertainty resulting from recent climate change is especially dangerous for those already living with a chronic water shortage, who have no reserves to fall back on.

DROUGHTS

The lives and livelihoods of one billion people, a sixth of the world's population, are threatened by droughts and desertification. Climate change is making the situation worse.

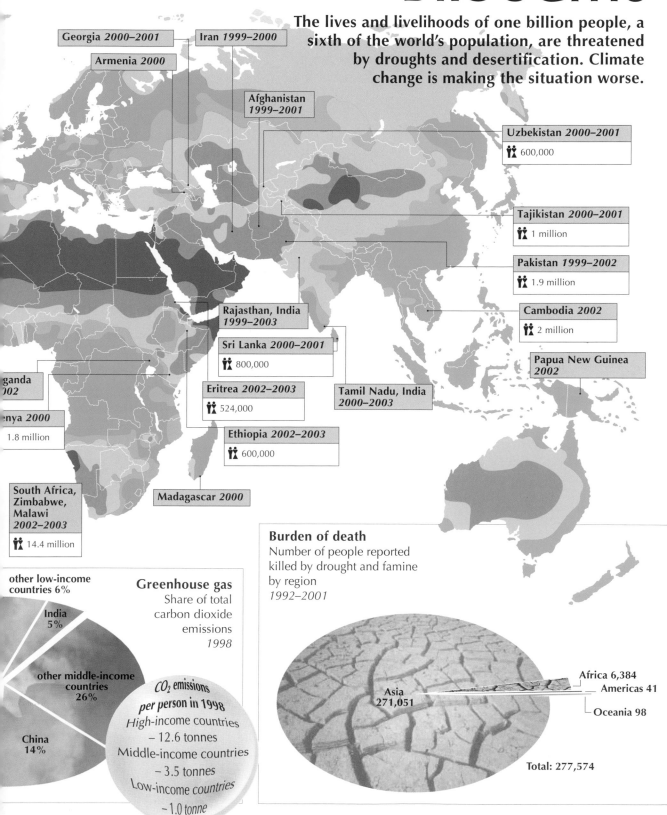

Georgia *2000–2001*

Iran *1999–2000*

Armenia *2000*

Afghanistan *1999–2001*

Uzbekistan *2000–2001*
👫 600,000

Tajikistan *2000–2001*
👫 1 million

Pakistan *1999–2002*
👫 1.9 million

Rajasthan, India *1999–2003*

Cambodia *2002*
👫 2 million

Sri Lanka *2000–2001*
👫 800,000

Papua New Guinea *2002*

...ganda ...02

Eritrea *2002–2003*
👫 524,000

Tamil Nadu, India *2000–2003*

...nya *2000*
1.8 million

Ethiopia *2002–2003*
👫 600,000

Madagascar *2000*

South Africa, Zimbabwe, Malawi *2002–2003*
👫 14.4 million

Greenhouse gas
Share of total carbon dioxide emissions *1998*

other low-income countries 6%

India 5%

other middle-income countries 26%

China 14%

CO₂ emissions per person in 1998
High-income countries – 12.6 tonnes
Middle-income countries – 3.5 tonnes
Low-income countries – 1.0 tonne

Burden of death
Number of people reported killed by drought and famine by region *1992–2001*

Asia 271,051

Africa 6,384

Americas 41

Oceania 98

Total: 277,574

PART 5

WATER CONFLICTS

THIS WORLD IS A TANGLED WEB OF WATERWAYS that ignores national boundaries. Along these waterways flourish the great centres of civilization, of which the first – it is sobering to remember – was that of the Tigris–Euphrates valley, where agriculture was first invented and irrigation first practised more than 6,000 years ago. This may also have been the site of the Great Flood of the Bible; indeed, an ancient Sumerian legend recounts the deeds of the deity Ea, who punished humanity for its sins by inflicting the Earth with a six-day storm.

If this is the first example of the vindictive use of water, it is by no means the last. The use of water as a military tool has a long and distinguished history. In 1503 Leonardo de Vinci and Machiavelli planned the diversion of the river Arno away from Pisa at a time when Florence and Pisa were warring states. Since then, levees were breached during the American Civil War, the Los Angeles pipeline was repeatedly bombed, and – in an attempt to repel the Japanese but in fact killing up to a million people in China – flood control dykes were breached on the Yellow River. Both German and Allied forces bombed dams, flooded marshes and created lakes during the Second World War in the pursuit of military victory.

Things have become more vicious. Israel, Jordan and Syria have battled since the 1950s for more effective control of their sparse water resources; the saga continues. During the 1990–91 Gulf War, Iraq wantonly destroyed much of Kuwait's desalination capacity during its retreat, having already poisoned and diverted the waters on which the Marsh Arabs of southern Iraq depended for their livelihoods and lifestyles. Dams and hydroelectric plants were bombed, and wells poisoned with dead bodies during the 1998–2000 Kosovo conflict.

Even where countries are not actively involved in military conflict, tensions about how to share scarce resources are present almost wherever rivers are shared. And many are. At least 260 river basins cross 145 international boundaries. Thirteen river basins are shared by five or more countries. Those who live upstream are the lucky ones; those who are downstream are vulnerable, and prone to maltreatment. At least 40 percent of the world's population lives in a river basin shared by two or more countries, and at least one-fifth of the world's population is under potential threat from upstream neighbours.

The good news here is that many countries have taken their riverine obligations seriously and set up commissions to manage shared waters. The Rhine, Zambezi, Nile, Mekong and Danube are among the best known. Formal cooperation is unexpectedly more advanced in the wet north than the dry south. While Europe has only four river basins shared by more than four countries, nearly 200 separate treaties govern the use of these waters.

The wars of this century will be fought over water

Conflict over water infiltrates at yet another level. Where water is short, who shall use it? Farmers, say the farmers, for we provide the food on which we must survive. Industry, says industrialists, for the use to which we put water is more productive than any other. Ordinary people, say the others, for if we don't have enough water in the home we can neither work in industry nor on the farm. All have their points, though those made by some industrialists and by some farmers are weaker than others. Few people, other than shareholders, can seriously support the actions of a soft drinks company that uses so much water that it drains the aquifer beneath a community of smallholders in India. In the south-west of the USA, fierce squabbles rage between farmers and cities as to who owns the water. Meanwhile, the once-mighty Colorado river has been reduced to a trickle, and a highly saline one at that, and downstream Mexico has been deprived of an important resource. Since the USA has treaty obligations to provide Mexico with at least 1.8 cubic kilometres of water a year, it has been forced to build a desalination plant to ensure that what water does reach Mexico is potentially drinkable.

Water dependency

Percentage of total renewable water resources originating outside country *2003*

- 76% – 100%
- 51% – 75%
- 26% – 50%
- 11% – 25%
- 10% or less
- no data

Co-operation

✓ countries that have signed or ratified the 1997 UN Convention on the Law of the Non-Navigational Uses of International Watercourses *2001*

Countries that draw water from the lower reaches of rivers or from shared aquifers depend on their neighbours' co-operation for the continuation of a good quality supply.

More than 260 river basins are international and 13 are shared by five or more countries. Disputes arise largely over how much water each country takes, and over the vexed issue of dams. Most of the violent conflicts have been between Israel and its neighbours.

The good news is that there have been more than twice as many positive interactions between countries over water as negative ones. Some water agreements have continued even in times of war. These include the Indus Water Commissions, established between India and Pakistan in 1960, and the Mekong River Committee, established in 1957 between Thailand, Cambodia, Vietnam and Laos.

In 1997, the UN passed a Convention on the Law of the Non-Navigational Uses of International Watercourses. Only three countries voted against it (China, Turkey and Burundi), but of the 103 that voted in favour, only 12 had signed or ratified it by 2001.

18 international agreements have been made relating to the waters of The Nile

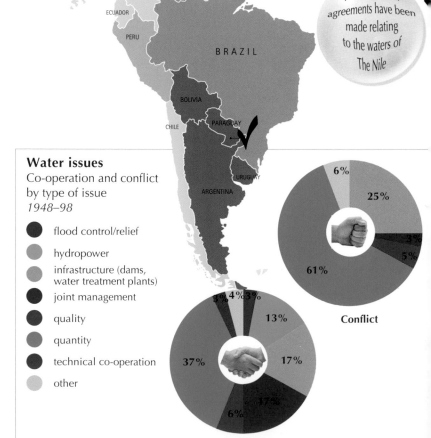

Water issues

Co-operation and conflict by type of issue *1948–98*

- flood control/relief
- hydropower
- infrastructure (dams, water treatment plants)
- joint management
- quality
- quantity
- technical co-operation
- other

Conflict

6%
25%
3%
5%
61%

Co-operation

3% 4% 3%
13%
17%
37%
6%
17%

THE NEED FOR CO-OPERATION

Sharing water supplies requires interaction between countries. In some cases the water cements friendships; in others it broadens rifts.

17 countries share the Danube's waters

SWEDEN
FINLAND
ESTONIA
LATVIA
RUSSIA
LITHUANIA
RK
POLAND
BELARUS
CZECH REPUBLIC
SLOVAKIA
UKRAINE
AUSTRIA
HUNGARY
MOLDOVA
OVENIA
ROMANIA
CROATIA
B-H
BULGARIA
LY
YUGOSLAVIA
ALBANIA
MACEDONIA
GREECE
MALTA

RUSSIA
KAZAKHSTAN
MONGOLIA
NORTH KOREA
JAPAN
SOUTH KOREA
CHINA

GEORGIA
ARMENIA
AZERBAIJAN
UZBEKISTAN
KYRGYZSTAN
TURKMENISTAN
TAJIKISTAN
IRAN
AFGHANISTAN
see inset
LIBYA
BAHRAIN
QATAR
UAE
PAKISTAN
OMAN
YEMEN
NEPAL
BHUTAN
BANGLADESH
INDIA
BURMA
LAOS
VIETNAM
THAILAND
CAMBODIA
TAIWAN
PHILIPPINES

NIGER
CHAD
SUDAN
ERITREA
DJIBOUTI
ETHIOPIA
SOMALIA
MALDIVES
SRI LANKA
BRUNEI
MALAYSIA
SINGAPORE

NIGERIA
CENTRAL AFRICAN REP.
CAMEROON
ATORIAL GUINEA
GABON
TOMÉ RINCIPE
CONGO
DEMOCRATIC REPUBLIC OF CONGO
UGANDA
KENYA
RWANDA
BURUNDI
TANZANIA
COMOROS
INDONESIA
PAPUA NEW GUINEA
SOLOMON ISLANDS

ANGOLA
ZAMBIA
MALAWI
MADAGASCAR
MAURITIUS
REUNION

NAMIBIA
ZIMBABWE
BOTSWANA
MOZAMBIQUE
SWAZILAND
SOUTH AFRICA
LESOTHO

AUSTRALIA
FIJI
NEW ZEALAND

Inset
TURKEY
CYPRUS
LEBANON
SYRIA
WEST BANK
GAZA STRIP
(ISRAEL)
IRAQ
JORDAN
KUWAIT
EGYPT
SAUDI ARABIA

Conflict and co-operation
Number and type of events related to transboundary river basins
1948–98

Event type	Number
intensive military act	21
small-scale military act	16
political/military hostile act	6
diplomatic/economic hostile act	50
strong/official verbal hostility	164
mild/unofficial verbal hostility	250
neutral, non-significant act	96
mild verbal support	430
official verbal support	190
cultural, scientific agreement/support	170
economic, technical, industrial agreement	296
military, economic strategic support	7
international water treaty	167

Central Asia *The demand for water in Central Asia is growing, and the competition for water is increasing political tension. The main sources of surface water, the Syr Darya and the Amu Darya, are largely controlled by Kyrgyzstan and Tajikistan. Disputes arise as Turkmenistan, Uzbekistan and Kazakhstan demand more water for irrigation. Uzbekistan and Kazakhstan trade coal and gas for summer water, but with deliveries unreliable, the upstream countries use their hydropower plants to generate electricity in winter, flooding their downstream neighbours.*

Several engineering schemes – including an artificial lake in the Karakum desert, and further dams in Tajikistan – are likely to ratchet up the political tension. The economic development of post-war Afghanistan is likely to place increasing demands on the river. There are even signs of imminent armed conflict. Uzbekistan has carried out military exercises that appear to rehearse the capture of the Toktogul Reservoir in Kyrgystan.

As populations grow and people congregate in cities, increasing demands are made on the world's finite freshwater resources. Conflict appears inevitable.

Disputes over river water are escalating, as downstream countries object to the plans of their upstream neighbours. Bangladesh has protested at a proposal by India to connect 37 rivers, including the Ganges and Brahmaputra rivers. Iraq is concerned that Turkey's plans to build more dams on the Euphrates river, and to divert water for irrigation, will reduce its own quota of river water by up to 90 percent.

At a local level, commercial operations that rely on plentiful supplies of water are competing with the surrounding community. In India, the villagers of Perumatty, Kerala, have complained that a large bottling plant owned by Coca Cola, which uses up to 4.5 million litres a day, is draining their water supplies.

Elsewhere, attempts to relieve water shortages in one region by bringing in water from another have led to court battles. Protestors in Alaska have defeated a commercial proposal to use water from the Gualala and Albion rivers to fill huge bags that would be towed down the coast to California.

USA and Mexico *The Colorado River is the main source of surface water for those living in its basin. The Lower Basin provides water for 17 million people, and irrigates more than 405,000 hectares, but the distribution of water between Arizona, California and Nevada has long been disputed. Since 1929, legislation has attempted to limit California's annual share to 5.4 km³, but the state overdraws, to provide water to the extensively irrigated Imperial Valley, and to its rapidly growing population.*

However, in 2003 four water companies in California signed up to a plan to reduce the amount of water for irrigation, and, over a 14-year period, to decrease the state's dependence on the river. Mexico was guaranteed 1.8 km³ of water a year in 1945, but in 1962 it protested to the USA about the salinity of the water it was left with. Since 1973, the USA has been responsible for ensuring the quality of the water flowing to Mexico, which has involved building and operating the Yuma Desalting Plant.

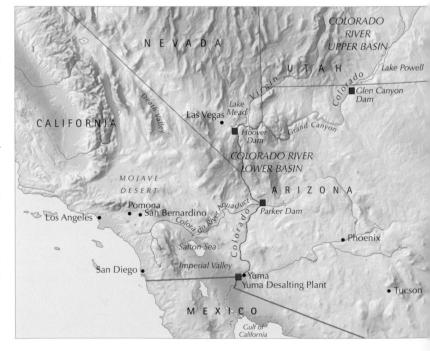

PRESSURE POINTS

Scarce water resources are increasing political tensions between and within countries, and between communities and commercial interests.

Israel and Palestine *In the Middle East water is a precious resource, and a source of conflict. The Six-Day War in 1967 was, in part, Israel's response to Jordan's proposal to divert the Jordan river for its own use. The land seized by Israel in the war gave it access not only to the headwaters of the Jordan, but to control of the aquifer beneath the West Bank, thereby increasing Israel's water resources by nearly 50%.*

Water is a major issue in Israeli-Palestinian negotiations. In the West Bank some Palestinians survive on as little as 35 litres a day for their domestic use, while Israeli settlers nearby enjoy lawns and swimming pools. Although the Declaration of Principles of 1993 led to a common Water Development Program, the refusal of Israel to acknowledge that scarce water resources should be equally shared is a major sticking point. Under the 1995 Oslo Interim Agreement, Israel retains overall control of the supply of water to the West Bank.

Israel extracts up to 75% of the flow of the Upper Jordan River, leaving only a brackish trickle to reach the West Bank. The National Water Carrier – 200 km of open canals, pipes and ducts completed in 1964 – transports 400 million m³ of water a year from the north to the more arid coastal regions.

Underground aquifers are also controlled by the Israeli government, including access to the mountain aquifer – the only source of water for the West Bank. Because the coastal aquifer has been over-used, it is being recharged from wastewater and by water brought by the National Water Carrier. In Gaza, over-use of the aquifer is leading to severe salinization of the water.

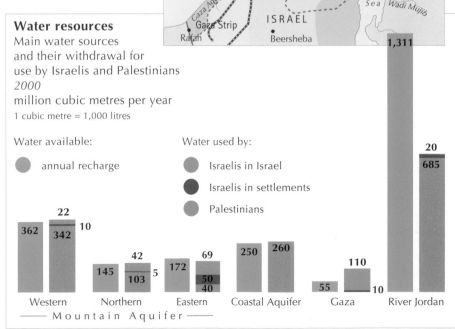

Israel and Palestine

- ▬ ▬ Israeli National Water Carrier
- Gaza Aquifer
- Coastal Aquifer
- Mountain Aquifer:
  - Northern Aquifer
  - Western Aquifer
  - Eastern Aquifer

Water imbalance

Average domestic water allowance per person per day *2002*

Palestinians
71 litres a day

Israelis
350 litres a day

Water resources

Main water sources and their withdrawal for use by Israelis and Palestinians *2000*
million cubic metres per year
1 cubic metre = 1,000 litres

Water available:
- ● annual recharge

Water used by:
- ● Israelis in Israel
- ● Israelis in settlements
- ● Palestinians

	Western	Northern	Eastern	Coastal Aquifer	Gaza	River Jordan
annual recharge	362	145	172	250	55	1,311
Israelis in Israel	342	103	50	260	—	685
Israelis in settlements	22	42	69	—	110	20
Palestinians	10	5	40	—	10	—

─── Mountain Aquifer ───

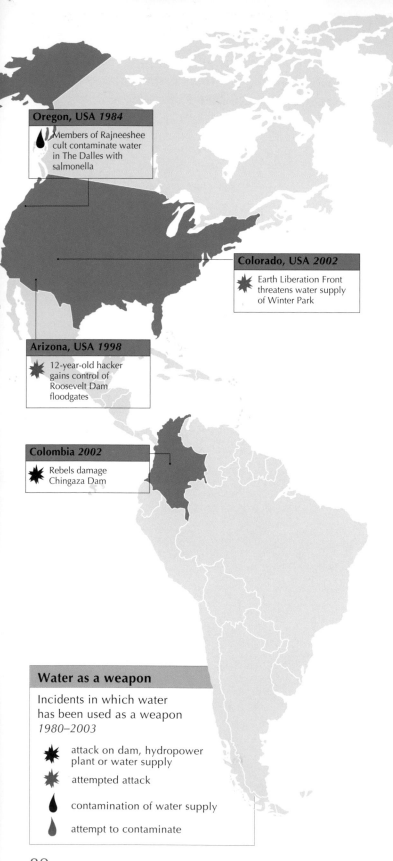

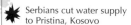 NATO planes target hydroelectric plant; shut down water supply

 Serbians cut water supply to Pristina, Kosovo

Serbs dump bodies in wells in Kosovo

Oregon, USA *1984*

Members of Rajneeshee cult contaminate water in The Dalles with salmonella

Croatia *1993*

 Peruca Dam attacked by Serbs

Italy *2002*

Terrorists' plan to contaminate Rome's water system is foiled

Colorado, USA *2002*

Earth Liberation Front threatens water supply of Winter Park

Bosnia-Herzegovina *1992*

 Bosnian Serbs cut water supply to Sarajevo

Arizona, USA *1998*

12-year-old hacker gains control of Roosevelt Dam floodgates

Macedonia *2001*

 Water flow cut to Kumanovo

Zambia *1999*

 Terrorist bomb cuts supply to Lusaka

Colombia *2002*

Rebels damage Chingaza Dam

Dem. Rep. Congo *1998*

 Inga Dam attacked by rebels

Angola *1999*

Bodies found in wells

Angola–Namibia *1988*

Cuban and Angolan forces attack Calueque Dam in Angola, threatening water supply to northern Namibia; pipeline to Owamboland destroyed

South Africa *1990*

 South African government cuts water to Wesselton township

Water as a weapon

Incidents in which water has been used as a weapon
1980–2003

 attack on dam, hydropower plant or water supply

 attempted attack

contamination of water supply

attempt to contaminate

WEAPON OF WAR

The deliberate destruction of dams and pipelines, and the contamination of drinking water are methods that have been used by both governments and terrorists against the military and civilian populations alike.

...ria 1990
- Turkey threatens to cut supply to Syria

Iraq
- **1980–88** Iran diverts water to flood Iraqi defence positions
  1981 Iran bombs dam in Kurdistan
  1991 Allied coalition targets Baghdad's water supply
- **1993** Saddam Hussein poisons and drains water supply of Marsh Arabs
- **2003** Main water pipeline to Baghdad is sabotaged by Saddam loyalists

Kazakhstan 2000
- Kyrgyzstan and Uzbekistan cut water supplies for non-payment of debts

Tajikistan 1998
- Guerrillas threaten to blow up dam

Lebanon 1982
- Israel cuts water supply to Beirut

Afghanistan 2001
- USA bombs hydroelectric plant near Kandahar

China 2000
- Officials in Guangdong province blow up water channel to prevent neighbouring county from diverting water

...rael–Palestine 2001
- Palestinians and Israelis disrupt each other's water supplies

Nepal 2002
- Maoist rebels destroy micro-hydro projects and damage water supplies in west

...wait 1991
- Iraq destroys Kuwait's desalination plants

East Timor 1999
- Opponents to independence dump bodies in wells

Singapore 1997
- Malaysia threatens Singapore's supplies

Australia 2000
- Man is arrested in Queensland for attempting to use computer and radio transmitter to take control of wastewater system and release sewage

81

PART 6

WAYS FORWARD

GOOD WATER MANAGEMENT is what decides the water fate of the most of the world's people. With few exceptions (Bahrain is one), aridity is only one aspect of a situation in which water availability combines with other factors – including ease of access to water, the efficiency with which resources are used, the capacity of people to benefit and the health of the environment, to determine what is known as 'water poverty'. The Water Poverty Index is a measure of integrated water management – the integration of decisions made about water with decisions made about land use and the environment, economic and social issues, and local, national and international requirements.

Much depends on whether water is seen as a human right or just as a human need which, like food and shelter, can be best provided for through market mechanisms. The Second World Water Forum in The Hague in March 2000 did little to help by defining water as a commodity. In this it was clearly hasty since, as Peter Gleitz points out, 'International law, declarations of government and international organizations, and state practices support the conclusion that access to a basic water requirement must be considered a basic human right'.

If access to a basic water requirement is a human right, then that right cannot be sold; fortunately, in most countries people do not have to buy the right not to be tortured or executed by government bodies. Neither should they have to buy the right to be given access to water. It is something they already own.

One compelling argument against privatization is this: who will buy water for nature? The value of wetlands and their wildlife is well known to everyone, except apparently private water companies. There is no profit for them in leaving part of what they regard as their private commodity for nature to do with as it sees fit. Nor are private companies much interested in providing water for people too poor to pay for it – though government subsidies can provide a neat fix that prevents the water companies from becoming poor

Fresh water is central to sustainable development, economic growth, social stability and poverty alleviation.
World Water Forum, 2000

as well. Even so, profiting from thirsty poor people is not a pretty concept.

It should not be necessary to say all this. In the kind of world in which we now live – one where commodity is the key concept and choice the altar at which politicians worship – the obvious has been forgotten or deliberately neglected by those who set our goals and develop strategies for achieving them.

Politicians' views are not universally shared. Fortunately, an army of ordinary mortals, mostly in developing countries and many of them women, have very different views. Their view is not a lofty one, seen from a penthouse office suite. It is the view from the window of a tapless kitchen of the track that leads to a muddy and infected water source five kilometres away. From Cochabamba in Bolivia to Grenoble in France, citizens groups are themselves taking on the mighty task of deprivatizing water companies. And they are succeeding. Participation is the name of this vital game.

They are succeeding, too, in finding out how to make the best use of the water resources that the sky provides. Thousands of years ago agriculturalists in the Negev desert knew how to channel gentle winter rains through freshly-planted fields to put the scarce resource to good use.

Their techniques are now being copied, not just in Israel but in many African countries as well. Spreading the good news about this and other cheap micro-techniques should become a major goal for our information society.

The World Water Council's detailed study of how our water world might look in the year 2025 makes it clear that a 'business as usual' approach will simply worsen the world disaster currently being acted out. Their rosiest vision of the future is attractive, though the assumptions on which it is based – including full water pricing, intensive research and much new technology – are more dubious. The two keys, as this publication emphasizes, are integrated water management and public participation. Using them we may yet see something of the IUCN's vision (*see page 91*).

If the world is to meet the UN's Millennium Goal of halving the proportion of the population without access to an improved water supply, and to sanitation, around US$150 billion needs to be spent on installing water pipelines, sewage systems, and wastewater treatment plants.

But where will the money come from? International aid contributes some of the costs, but donor countries are not giving nearly enough to cover the work needed. Increasingly, governments are granting concessions to private companies, allowing them to run the water system as a commercial enterprise. While this policy is now well established in richer economies, it is a dangerous precedent to entrust profit-led organizations with the provision of such a vital resource to the world's poorest people.

Neither does privatization solve the problem of developing and maintaining infrastructure. In general, commercial organizations on time-limited concessions are unwilling to make hefty investments, preferring to go for a quick profit. So in order to encourage firms to improve and maintain infrastructure, governments are forced to grant the companies generous terms, including guarantees of profit. However, the Bolivian government's decision to privatize the water sources, as well as the supply system, proved a step too far, and sparked civil unrest during 2000. The price of water in Cochabamba tripled, and the rural poor were forced to pay for water they considered theirs by right. By the end of the year, the government had been forced to back out of the deal, and the company was suing for a broken contract.

The market in bottled water was estimated to be worth more than US$20 billion a year in 2002, and is expanding rapidly. A number of factors account for this: distaste for tap water and fears over its safety, plus a growing realization that water is healthier than other bottled or canned drinks.

Bottled water
Average consumption per person per year
2002 or latest available

101 litres — West Europe
75 litres — USA
20 litres — East Europe
7 litres — Asia

Asia | East Europe | West Europe | USA

Sales of bottled water in the USA leapt by 11% in one year between 2000 and 2001

Big business
The World Bank judged the global trade in water to be US$1,000 trillion in 2001

Wages for water
Percentage of wage spent on water
2003
selected countries

USA 0.006%
UK 0.013%
Pakistan 1.1%
Uganda 3.2%
Tanzania 5.7%

Who pays what?
Price paid by businesses in OECD countries for 1,000 litres of water
2001
US$
selected countries

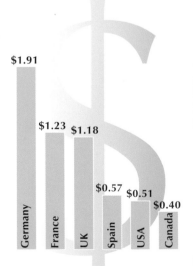

Germany $1.91
France $1.23
UK $1.18
Spain $0.57
USA $0.51
Canada $0.40

THE WATER BUSINESS

Water is a vital natural resource, but getting it to those who need it is a problem for governments. For commercial enterprises, though, it is a major opportunity.

Water globalization
Number of people the Suez company serves with water and sanitation *2003*

NORTH AMERICA **23.5m**

EUROPE **43.0m**

ASIA and PACIFIC **25.0m**

SOUTH AMERICA **25.0m**

AFRICA **8.5m**

Water supply is a huge growth area, and is increasingly attracting multinational companies. One such, Suez, claims to serve 125 million people per day with water and sanitation services.

The poor pay more
Price paid for 1 cubic metre of water supplied through a pipe or in a container
1997
selected cities
US$
1 cubic metre = 1,000 litres

What price water?
The price of water in the developed world varies according to availability, subsidy, and commercial arrangements. In general, the price paid by the consumer has a direct influence on the amount that is used, and pricing tariffs based on consumption discourage waste.

Subsidies for agricultural water use often lead to waste, and to inappropriate crops. In Spain, for example, the price paid by some farmers is only 1% of the true cost of providing the water, which is why they are able to grow water-guzzling plants such as maize, alfalfa and potatoes, in some of the

driest land in Europe. In the western USA, government water subsidies to farmers total more than US$2 billion a year.

When it comes to paying for water, the poorest, and those who use the least, pay the highest price. People in developing countries often have to spend a much larger proportion of their income on ensuring a supply of water, than do those in richer countries.

People who cannot afford to have water piped to their home, have to resort to unregulated water vendors selling water by the container. They have little choice but to pay what is demanded of them.

household connection		informal vendor
$0.08	Dhaka, Bangladesh	$0.42
$0.16	Jakarta, Indonesia	$0.31
$0.14	Karachi, Pakistan	$0.81
$0.11	Manila, Philippines	$4.74
$0.09	Phnom Penh, Cambodia	$1.64
$0.04	Ulaanbaatar, Mongolia	$1.51

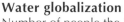

There is a pressing need to conserve the world's water supplies, made more urgent by the uncertainty over future climate trends.

There are a number of ways water use can be improved. One is to increase the amount of water available by 'capturing' more of it. Another is to waste less water, or to reduce demand, so that less water needs to be withdrawn. The other is to use the same amount of water, but in a more effective way.

'Water harvesting' involves collecting and using rainwater that would otherwise soak away or evaporate without doing any good. In many regions this is part of a centuries-old tradition, but one that may need reviving and disseminating more widely. Water harvesting can be extraordinarily effective. Water collected from village roofs during the monsoon rains and stored underground, can make the difference between crop failure and a successful harvest.

Repairing underground pipes, using more economic methods of irrigation, and installing water-saving appliances in the home, are all ways of reducing the amount of water needed.

Water can be used more effectively in a number of ways. Improved agricultural practices are producing 'more crop per drop', and treated wastewater is being used to irrigate crops. Many modern industries require less water than previously. But gains in one area may have a negative impact on another, such as when agricultural run-off destroys fish stocks downriver or pollutes drinking water.

Frankfurt airport terminal captures 16 million litres of rain a year on its roof for use in cleaning, gardening and toilets

INDUSTRY

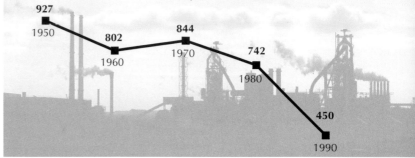

Industrial savings
Industrial water use per person per day in the USA
1950–1990
litres

927 1950
802 1960
844 1970
742 1980
450 1990

Industries in the developed countries, driven by regulations and the desire to cut costs, have generally reduced the amount of water they use. Steel can now be produced using less than a quarter of the water it once used. In the USA, industrial use per person halved between 1950 and 1990, while industrial output nearly quadrupled. Unless the newly industrializing countries learn the water conservation lesson, they will place unacceptable demands on water resources.

DOMESTIC

About 30% of water used in the home in developed countries is flushed down the toilet (see page 30), so low-flush toilets make a significant contribution to conserving water. A third of California's urban water use could be saved by installing water-saving technologies, which would be cheaper than tapping into new sources of supply.

In Mexico City, regulations introduced in 1989, and an increase in water charges, led to 350,000 low-flush toilets being installed, and a saving of 28 million cubic metres of water a year.

Conventional toilet **16 litres per flush**

Low-flush toilet **6 litres per flush**

Wastewater reuse
Uses of wastewater in Mexico City
2000
percentages

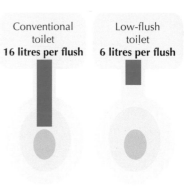

refilling of Texcoco Lake *(primary and secondary treatment)* **2%**

animal drinking water and groundwater recharge *(tertiary treatment)* **1% or less**

irrigation of green spaces in city *(primary treatment)* **22%**

irrigation in Hidalgo State *(untreated)* **75%**

In many water-short countries, city sewers play a vital role in irrigating crops. In some cases the water is first purified. In Calcutta, India, raw sewage is channelled into natural lagoons, which soak up and clean up the sewage, and produce 6,000 tonnes of fish a year. In Israel, more high-tech treatment plants are used on wastewater, which is used to irrigate 20,000 hectares of farmland.

Elsewhere, wastewater is used untreated. This is a relatively cheap way of increasing crop production, but it does bring health risks. The Mexico City authorities stipulate that untreated wastewater can only be used on grains and animal fodder, but that is not the approach adopted everywhere.

CONSERVING SUPPLIES

Fresh water is an increasingly scarce and valuable resource; it should be used in the most effective way possible.

AGRICULTURE

Methods of irrigation
Percentage of irrigation by method
2003
selected countries

- surface irrigation
- sprinkler irrigation
- drip irrigation

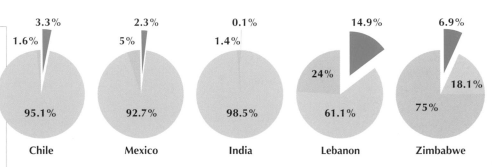

	Chile	Mexico	India	Lebanon	Zimbabwe
	3.3%	2.3%	0.1%	14.9%	6.9%
	1.6%	5%	1.4%	24%	18.1%
	95.1%	92.7%	98.5%	61.1%	75%

A third of all vegetables grown in Asmara, Eritrea, are watered with wastewater

Surface irrigation involves water being directed into channels between and around planted areas. In **sprinkler irrigation**, water is sprayed onto a field. In both of these methods, up to 25% of the water may be lost through evaporation.

In **drip irrigation**, water is fed through hoses or pipes under pressure, and drips through holes or nozzles. Loss through evaporation is just 5%, and because water can be directed to the roots of plants, it usually leads to increased yields. Despite its obvious advantages, the cost of the equipment needed means that drip irrigation is still not widely practised, with the exception of Israel, where it is used on 50% of irrigated land.

Less water, more rice
Effect of reduced irrigation levels on rice yield per hectare
1992–1993

Water harvesting
Water collected by 70 villages in Rajasthan caused the once-seasonal Arvari river to flow all year round

The conventional method for growing rice involves keeping the paddy field continually under water. However, experiments in Guangxi province, China, proved that at some stages of its growth, paddy actually benefits from lower water levels, and researchers were able to grow more rice using less water.

5,907 m³   5,866 kg   4,659 m³   6,534 kg

1992
average water use — average rice yield

1993
average water use — average rice yield

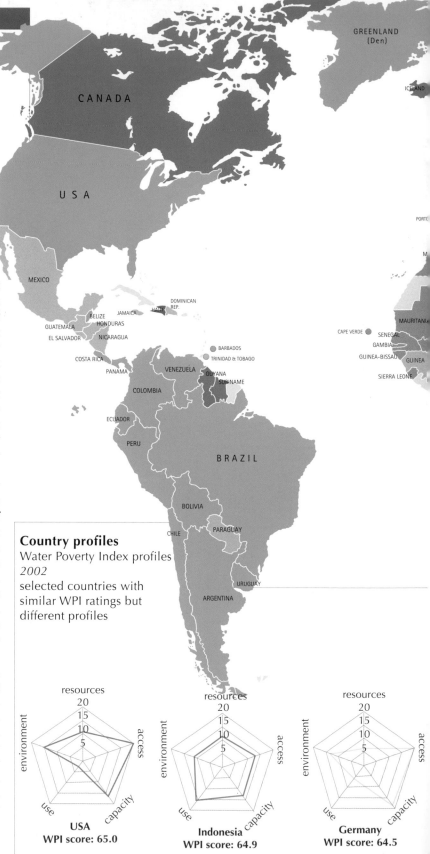

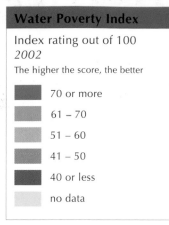

Water Poverty Index

Index rating out of 100
2002

The higher the score, the better

- 70 or more
- 61 – 70
- 51 – 60
- 41 – 50
- 40 or less
- no data

Integrated water management is generally recognized as key to dealing with water scarcity.

When devising a management strategy, the physical resource is often the starting point. Satellite images and water accounting are used to determine how much water there is, and how productively it is used. But the way the water is used and managed within a community, and its impact on the surrounding environment, also need to be taken into account. Only then can the link between poverty, and poor water and sanitation be properly understood and addressed.

Those concerned to manage water resources in an integrated way need to ask questions about a wide range of issues. The Water Poverty Index provides a framework for making such an assessment, and can be used to monitor national water management. But a single country score hides wide variations. The index is a tool with which small communities can assess their own needs and manage their water use.

Truly integrated management is a bottom-up approach. Pieces of the jigsaw that have often been missing, need to be identified and incorporated into strategy at both a local and national level. This includes, for example, the important role played by women in developing countries in providing water for their families and managing the water supply for their community.

Country profiles

Water Poverty Index profiles
2002
selected countries with similar WPI ratings but different profiles

USA
WPI score: 65.0

Indonesia
WPI score: 64.9

Germany
WPI score: 64.5

SETTING PRIORITIES

Water resources need to be managed in an integrated way – by addressing people's social, economic, and health needs alongside the needs of the environment.

RUSSIA

NORWAY
SWEDEN
FINLAND
DENMARK
NETH.
GERMANY POLAND BELARUS
CZECH REP.
SWITZ. AUSTRIA HUNGARY MOLDOVA
SLOVENIA
CROATIA ROMANIA
ITALY BULGARIA
GEORGIA
GREECE ARMENIA
TURKEY AZERBAIJAN
CYPRUS SYRIA
LEBANON
ISRAEL JORDAN
TUNISIA

KAZAKHSTAN

MONGOLIA

UZBEKISTAN KYRGYZSTAN
TURKMENISTAN TAJIKISTAN
CHINA

SOUTH KOREA JAPAN

KUWAIT
BAHRAIN
QATAR
UAE
EGYPT SAUDI ARABIA
IRAN
PAKISTAN
NEPAL BHUTAN

INDIA BANGLADESH
OMAN BURMA LAOS
NIGER CHAD ERITREA YEMEN THAILAND VIETNAM
SUDAN DJIBOUTI CAMBODIA
NIGERIA ETHIOPIA
SRI LANKA PHILIPPINES
CAMEROON CENTRAL AFRICAN REP.
EQUATORIAL GUINEA UGANDA
GABON KENYA
CONGO RWANDA
BURUNDI
DEMOCRATIC REPUBLIC OF CONGO MALAYSIA
SINGAPORE
TANZANIA
COMOROS
ANGOLA
ZAMBIA MALAWI INDONESIA
PAPUA NEW GUINEA
ZIMBABWE
NAMIBIA BOTSWANA MADAGASCAR
MOZAMBIQUE MAURITIUS FIJI

SWAZILAND
SOUTH AFRICA LESOTHO

AUSTRALIA

NEW ZEALAND

A new well in Malica, Mozambique, was planned with full community involvement, supported by WaterAid. The villagers opted for an easy-to-maintain bucket and windlass system. They chose a site outside the village, where the water supply was reliable. Eight state-dug wells with handpumps, sited in the village without consulting the villagers, had proved unsustainable because of the problem of getting spare parts, and because they often ran dry.

A country's index rating is the sum of five scores out of 20, in which diferent aspects of water management are assessed:
Resources – amount of water available (renewable water supply)
Access – access to an improved water supply and to sanitation; irrigated land as a proportion of cropland and water resources

Capacity – GDP per capita, under-five mortality rate, school enrolment rates, degree of economic equality
Use – amount of water used per person (50 litres per day being considered an appropriate amount), the amount of water used for a sector (domestic, agriculture, industry) in proportion to the

GDP generated by that sector
Environment – water quality and stress, and the importance attached to water and environment.

Marked differences can be seen in the profiles of countries with similar index ratings. A country such as the USA, for example, may score highly on access, but poorly on use.

resources
20
15
10
5
environment   access

use   capacity

Equatorial Guinea
WPI score: 67.7

U nless radical steps are taken to alter the way water is withdrawn, used and managed, the outlook is bleak. By 2025, the world could be faced with a severe water shortage. This is likely to result in a significant reduction in food production, leading to malnutrition and disease, and to major ecological damage.

This gloomy scenario can be avoided. Widespread recognition of the problem, coupled with the desire to act, is a vital first step. Water management that looks at the whole picture, and involves the communities using the water, is essential, as is funding for research to develop water-conserving technologies, and co-operation between countries sharing river basins.

The most controversial measure proposed at the World Water Forum in 2000, was to charge people for the water they use. Advocates argue that this encourages water thrift; others point out that water is a human right to which people should have access regardless of whether they can pay.

Another controversial issue is the trade in 'virtual water', whereby rich, water-stressed countries, purchase other countries' water in the form of food. Some see this as a partial solution to water shortage, but others see it as exacerbating the problem, arguing that the rural poor in both the food-exporting and food-importing countries are unlikely to benefit from the trade.

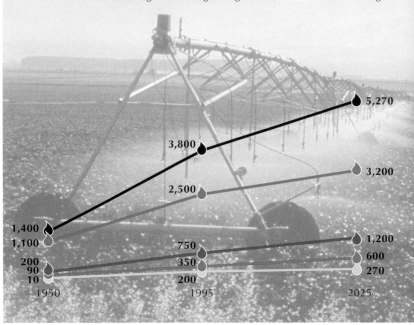

Scenario 1: Business as usual
Based on current population projections, this scenario assumes that the area of land being irrigated will continue to expand, leading to a 39% increase in total water withdrawals from 1995 to 2025. It takes no account of social, economic or political factors, and assumes that water usage and management practices, including the building of large dams, will continue unchanged.

5,270
3,800
3,200
2,500
1,400
1,100
750
1,200
600
200
90
10
350
200
270
1950   1995   2025

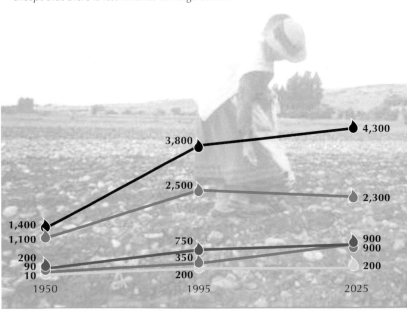

Scenario 2: Business nearly as usual
Based on current population predictions, this scenario assumes only a limited expansion of irrigated land area, leading to a chronic food shortage. The industrialization of developing countries results in increased domestic and industrial water use. Water management practices remain largely unchanged, except that there is less reliance on large dams.

4,300
3,800
2,500
2,300
1,400
1,100
750
900
900
200
90
10
350
200
200
1950   1995   2025

VISION OF THE FUTURE

The future of the world's water hangs in the balance.
Predicted scenarios vary, and depend on local,
national, and international policies and actions.

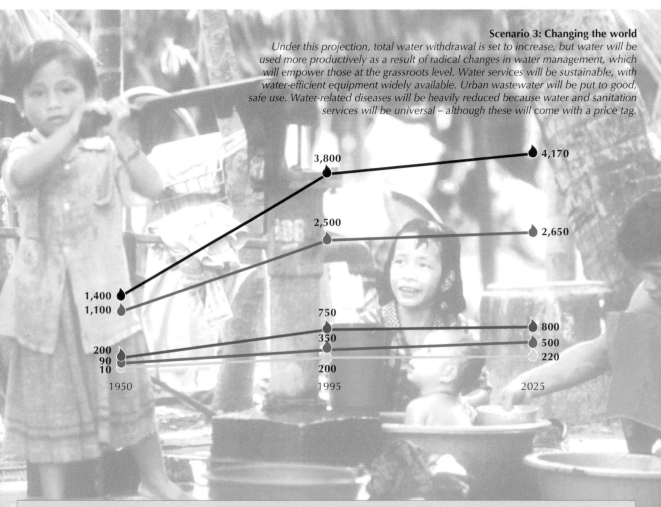

Scenario 3: Changing the world
*Under this projection, total water withdrawal is set to increase, but water will be
used more productively as a result of radical changes in water management, which
will empower those at the grassroots level. Water services will be sustainable, with
water-efficient equipment widely available. Urban wastewater will be put to good,
safe use. Water-related diseases will be heavily reduced because water and sanitation
services will be universal – although these will come with a price tag.*

3,800   4,170

2,500   2,650

1,400
1,100

750   800
350   500

200   220
90
10   200

1950   1995   2025

"Early on an April morning in 2025, on one of the vast floodplains of the Sahel, Ibrahim Diaw leads his herd of long-horned cattle to their dry-season pastures. The grazing routes for nomadic herders follow the areas under an ecosystem restoration programme initiated at the turn of the century. Using these migration pathways no longer results in violent conflicts with farmers, as it did 40 years before when intensive irrigated rice schemes were constructed throughout the plain. Now Diaw's herd prospers through access to large expanses of restored perennial grasses …

Throughout the wet and dry seasons, water holes provide drinking water for his animals, and the floodplain 'works' for the benefit of Diaw and other local people, who can count on stable livelihoods based on recession agriculture, semi-intensive production, and artisanal and small-scale commercial fishing. Diaw walks in the grass and thinks of the past – desiccated flats, 25 years without a single wedding in the village, his father who thought that they had been forgotten by God … Efforts to mitigate the impacts of infrastructure development are about to pay off: the dikes have been put to good use, artificial flooding schemes are effective, and water is not wasted anymore."

World Water Vision, 2000

OUR WORLD IS OBSESSED WITH NUMBERS. In some way we feel that if an issue is to be understood as important, it must have numbers attached. It must be responsible for X million deaths or Y thousand lost jobs or Z hundred disasters. This should not be so. Watching one child die of famine under a tree in a remote African settlement means more to those who witness it than all the statistics they ever could collect in a lifetime.

This obsession with numbers – the numeric fallacy – is responsible for many ills. It means that issues that are hard to quantify are often not given the importance they deserve. This book is no exception. Many subjects would have been dealt with here in more detail had there been more or better data to support them, or maps that could be used to illustrate them. The importance of participation and the role of women in decision making about water matters are two such subjects. The effects of climate change on our water future is another.

The numeric fallacy would be unhealthy even if the data used to quantify such things were reliable. They are not. Water data are particularly unreliable and the subject of prolonged and intense debate. During the preparation of this book we more than once posed questions about

Millennium Development Goal
Halve the proportion of people without access to safe water by 2015

startling contrasts in data from neighbouring places. There were no easy explanations except the obvious one: on a hot afternoon, a clerk sat at his desk trying to fill in a long form for his local Ministry. Deprived of any real data and bored to tears with the whole exercise, he entered the first figure that came into his head and looked at all reasonable. From here these 'data' went to the Ministry, where they were combined with others for more detailed analysis by an independent think-tank. Their report was published and the data it contained found its way into the United Nations' records. So it must be true, as they say.

We do not know how many 'data' were created in this way, but this is certainly a problem that needs investigation. That said, it would be even more foolish to follow the advice of some specialists in the area. "Don't publish an atlas of water", they said. "The data are not good enough to support it".

This is another numeric fallacy. If the data are really that inadequate, they should not be published. If they are published and available, then they are there to be used. The important thing is to use them advisedly and to warn readers that all that is made of numbers is not gold. Here is that warning.

Countries	1 Total population millions		2 Urban population		3 Improved water source % of population with access 2000	4 Improved sanitation % of population with access 2000
	2000	2050 projected	as % of total population 2000	annual % increase 2000–2005 projected		
Afghanistan	21.8	61.0	22%	5.7%	13%	12%
Albania	3.1	4.3	42%	2.1%	97%	91%
Algeria	30.3	57.7	57%	2.7%	89%	92%
Angola	13.1	36.9	34%	4.8%	38%	44%
Argentina	37.0	54.5	88%	1.4%	–	–
Armenia	3.8	4.0	67%	0.2%	–	–
Australia	19.1	25.8	91%	1.4%	100%	100%
Austria	8.1	7.1	67%	0.2%	100%	100%
Azerbaijan	8.0	10.0	52%	0.6%	78%	81%
Bahamas	0.5	0.5	89%	1.6%	97%	100%
Bahrain	0.6	1.0	92%	2.0%	–	–
Bangladesh	137.4	212.5	25%	4.3%	97%	48%
Belarus	10.2	8.3	69%	−0.2%	100%	–
Belgium	10.2	8.9	97%	0.2%	–	–
Belize	0.2	0.5	48%	2.2%	92%	50%
Benin	6.3	15.6	42%	4.5%	63%	23%
Bhutan	2.1	5.7	7%	5.9%	62%	70%
Bolivia	8.3	17.0	62%	3.0%	83%	70%
Bosnia-Herzegovina	4.0	3.8	43%	2.2%	–	–
Botswana	1.5	2.8	49%	1.4%	95%	66%
Brazil	170.4	244.2	81%	1.9%	87%	76%
Brunei	0.3	0.5	72%	2.5%	–	–
Bulgaria	7.9	5.7	67%	−0.9%	100%	100%
Burkina Faso	11.5	35.5	17%	5.1%	42%	29%
Burma	47.7	64.9	28%	2.9%	72%	64%
Burundi	6.4	15.6	9%	6.4%	78%	88%
Cambodia	13.1	20.7	17%	5.5%	30%	17%
Cameroon	14.9	37.3	49%	3.6%	58%	79%
Canada	30.8	42.3	79%	1.1%	100%	100%
Central African Rep.	3.7	7.7	41%	2.8%	70%	25%
Chad	7.9	19.7	24%	4.7%	27%	29%
Chile	15.2	22.2	86%	1.5%	93%	96%
China	1,282.0	1,484.4	36%	3.2%	75%	40%
Colombia	42.1	71.6	75%	2.3%	91%	86%
Comoros	0.7	1.6	33%	4.6%	96%	98%
Congo	3.0	8.6	65%	4.0%	51%	–
Congo, Dem. Rep.	50.9	160.4	30%	4.9%	45%	21%
Costa Rica	4.0	7.2	59%	2.9%	95%	93%
Côte d'Ivoire	16.0	30.5	44%	3.0%	81%	52%
Croatia	4.7	3.7	58%	0.8%	–	–
Cuba	11.2	11.1	75%	0.5%	91%	98%
Cyprus	0.8	0.9	70%	1.2%	100%	100%

Sources: Col 1: UN Population Division; **Col 2:** UN Millennium Indicator – UNICEF; UN Population Division; **Cols 3 & 4:** UNICEF;

NEEDS AND RESOURCES

5 Internal renewable water resources		6 Water dependency % of renewable water originating outside country 2000	7 Annual groundwater withdrawals m³ per person 2000	8 Desalination capacity m³ per day 1996	9 Water Poverty Index rating out of 100 2003	Countries
m³ per person per year 2000	m³ per person per year 2050 projected					
2,527	761	15%	–	–	–	Afghanistan
14,198	11,396	35%	194	–	–	Albania
459	272	3%	117	190,837	50	Algeria
14,009	3,450	0%	–	–	41	Angola
7,453	5,062	66%	180	15,960	61	Argentina
2,395	2,880	14%	–	–	54	Armenia
17,923	12,942	0%	143	82,129	62	Australia
6,968	8,726	29%	173	14,540	75	Austria
1,009		73%	–	12,680	–	Azerbaijan
		0%	–	37,474	–	Bahamas
6	4	97%	–	282,955	56	Bahrain
764	396	91%	98	–	54	Bangladesh
3,652	4,479	36%	116	12,640	61	Belarus
820	877	34%	79	3,900	61	Belgium
70,695	40,816	14%	–	757	66	Belize
1,642	570	58%	–	–	39	Benin
45,561	17,059	0%	–	–	56	Bhutan
36,444	17,891	51%	–	–	63	Bolivia
8,926	10,266	5%	–	–	–	Bosnia-Herzegovina
1,882	1,375	80%	–	–	57	Botswana
31,795	21,914	34%	57	1,079	61	Brazil
25,891	15,044	0%	–	–	–	Brunei
2,264	3,973	1%	566	1,320	63	Bulgaria
1,517	378	0%	–	–	42	Burkina Faso
18,442	12,847	16%	–	–	54	Burma
566	178	0%	–	–	40	Burundi
9,201	4,035	75%	–	–	46	Cambodia
18,016	8,301	4%	–	–	54	Cameroon
92,646	70,520	2%	37	35,629	78	Canada
37,931	17,206	2%	–	–	44	Central African Rep.
1,902	541	65%	16	–	39	Chad
58,115	38,888	4%	–	83,509	69	Chile
2,194	1,924	1%	47	–	51	China
50,161	29,804	1%	–	7,165	66	Colombia
1,417	526	0%	–	–	44	Comoros
73,548	20,663	30%	–	–	57	Congo
18,352	4,594	73%	–	–	46	Congo, Dem. Rep.
27,936	15,622	0%	–	–	67	Costa Rica
4,790	2,383	5%	–	–	46	Côte d'Ivoire
13,194	14,693	64%	–	–	68	Croatia
3,404	3,541	0%	408	18,926	–	Cuba
1,148	989	0%	–	6,275	62	Cyprus

Cols 5, 6, & 7: FAO Aquastat 2003; Col 8: Pacific Institute <www.worldwater.org/table16.html>; Col 9: Peter Lawrence, Jeremy Meigh and Caroline Sullivan: The Water Poverty Index: an International Comparison, Keele Economic Research Papers 2002/19, March 2003

Countries	1 Total population millions		2 Urban population		3 Improved water source	4 Improved sanitation
	2000	2050 projected	as % of total population 2000	annual % increase 2000–2005 projected	% of population with access 2000	% of population with access 2000
Czech Republic	10.3	7.8	75%	0.0%	–	–
Denmark	5.3	4.8	85%	0.2%	100%	–
Djibouti	0.6	1.3	84%	1.3%	100%	91%
Dominican Republic	8.4	12.3	65%	2.4%	86%	67%
Ecuador	12.6	21.2	63%	2.4%	85%	86%
Egypt	67.9	114.8	43%	1.8%	97%	98%
El Salvador	6.3	11.2	60%	3.5%	77%	82%
Equatorial Guinea	0.5	1.1	48%	4.9%	44%	53%
Eritrea	3.7	9.1	19%	6.3%	46%	13%
Estonia	1.4	0.9	69%	–1.1%	–	–
Ethiopia	62.9	169.4	16%	4.6%	24%	12%
Fiji	0.8	1.3	49%	2.5%	47%	43%
Finland	5.2	4.9	59%	0.1%	100%	100%
France	59.2	59.9	75%	0.6%	–	–
Gabon	1.2	2.7	81%	3.4%	86%	53%
Gambia	1.3	2.8	31%	4.4%	62%	37%
Georgia	5.3	5.2	56%	–0.1%	79%	100%
Germany	82.0	73.3	88%	0.2%	–	–
Ghana	19.3	51.8	36%	3.1%	73%	72%
Greece	10.6	8.2	60%	0.5%	–	–
Guatemala	11.4	27.2	40%	3.4%	92%	81%
Guinea	8.2	16.3	27%	3.1%	48%	58%
Guinea-Bissau	1.2	2.7	32%	4.8%	56%	56%
Guyana	0.8	1.2	36%	1.4%	94%	87%
Haiti	8.1	15.2	36%	3.3%	46%	28%
Honduras	6.4	13.9	53%	4.0%	88%	75%
Hungary	10.0	7.5	65%	–0.1%	99%	99%
Iceland	0.3	0.3	92%	0.8%	–	–
India	1,008.9	1,528.9	28%	2.3%	84%	28%
Indonesia	212.1	311.9	41%	3.6%	78%	55%
Iran	70.3	114.9	64%	2.4%	92%	83%
Iraq	22.9	54.9	68%	2.7%	85%	79%
Ireland	3.8	4.7	59%	1.4%	–	–
Israel	6.0	9.4	92%	2.2%	–	–
Italy	57.5	41.2	67%	0.1%	–	–
Jamaica	2.6	3.8	56%	1.8%	92%	99%
Japan	127.1	104.9	79%	0.4%	–	–
Jordan	4.9	16.5	79%	3.0%	96%	99%
Kazakhstan	16.2	18.7	56%	–0.3%	91%	99%
Kenya	30.7	51.0	33%	4.6%	57%	87%
Korea (North)	22.3	30.8	60%	1.2%	100%	99%
Korea (South)	46.7	51.3	82%	1.3%	92%	63%

Sources: Col 1: UN Population Division; **Col 2:** UN Millennium Indicator – UNICEF; UN Population Division; **Cols 3 & 4:** UNICEF;

5 Internal renewable water resources		6 Water dependency	7 Annual groundwater withdrawals	8 Desalination capacity	9 Water Poverty Index	Countries
m³ per person per year 2000	m³ per person per year 2050 projected	% of renewable water originating outside country 2000	m³ per person 2000	m³ per day 1996	rating out of 100 2003	
5,667	6,906	0%	48	11,085	61	Czech Republic
2,068	2,165	0%	170	5,960	61	Denmark
475	281	0%	–	–	38	Djibouti
2,508	1,768	0%	–	1,135	59	Dominican Republic
34,162	23,072	0%	–	4,433	67	Ecuador
27	16	97%	85	102,051	58	Egypt
2,831	1,637	30%	–	–	56	El Salvador
65,688	21,771	0%	–	–	68	Equatorial Guinea
765	279	56%	–	–	37	Eritrea
9,123	16,904	1%	–	–	62	Estonia
1,749	590	0%	–	–	35	Ethiopia
35,091	31,168	0%	–	–	62	Fiji
21,269	23,439	3%	48	–	78	Finland
2,870	2,749	12%	104	29,112	68	France
133,324	51,833	0%	1	–	62	Gabon
2,303	1,152	63%	–	–	48	Gambia
11,047	18,058	8%	550	–	60	Georgia
1,170	1,356	31%	89	–	65	Germany
1,569	756	43%	–	–	45	Ghana
4,260	5,032	22%	196	35,620	66	Greece
9,591	4,113	2%	–	–	59	Guatemala
27,716	10,912	0%	–	–	52	Guinea
13,342	4,884	48%	–	–	48	Guinea-Bissau
316,891	478,175	0%	–	–	76	Guyana
1,598	930	7%	–	–	35	Haiti
14,950	7,468	0%	–	651	60	Honduras
602	801	94%	97	500	61	Hungary
601,523	504,505	0%	559	–	77	Iceland
1,249	802	34%	223	115,509	53	India
13,381	9,116	0%	–	103,244	65	Indonesia
1,827	1,058	7%	739	423,427	60	Iran
1,534	657	53%	13	324,476	–	Iraq
12,358	8,759	6%	62	2,725	73	Ireland
281	169	55%	205	90,378	54	Israel
2,771	3,710	5%	243	483,668	61	Italy
3,651	2,465	0%	108	6,094	58	Jamaica
3,383	3,937	0%	101	637,900	65	Japan
138	58	23%	144	7,131	46	Jordan
4,663	4,929	31%	–	167,379	58	Kazakhstan
659	365	33%	–	–	47	Kenya
3,009	2,390	13%	–	–	–	Korea (North)
1,387	1,258	7%	55	265,957	62	Korea (South)

Cols 5, 6, & 7: FAO Aquastat 2003; Col 8: Pacific Institute <www.worldwater.org/table16.html>; Col 9: Peter Lawrence, Jeremy Meigh and Caroline Sullivan:
The Water Poverty Index: an International Comparison, Keele Economic Research Papers 2002/19, March 2003

Countries	1 Total population millions		2 Urban population		3 Improved water source % of population with access 2000	4 Improved sanitation % of population with access 2000
	2000	2050 projected	as % of total population 2000	annual % increase 2000–2005 projected		
Kuwait	1.9	3.5	96%	2.6%	–	–
Kyrgyzstan	4.9	7.4	34%	1.2%	77%	100%
Laos	5.3	13.3	19%	4.6%	37%	30%
Latvia	2.4	1.6	60%	−0.6%	–	–
Lebanon	3.5	5.2	90%	1.9%	100%	99%
Lesotho	2.0	4.8	28%	3.4%	78%	49%
Liberia	2.9	10.0	45%	6.8%	–	–
Libya	5.3	11.0	88%	2.5%	72%	97%
Lithuania	3.7	3.0	69%	0.0%	–	–
Luxembourg	0.4	0.4	92%	1.6%	–	–
Macedonia	2.0	2.3	59%	0.4%	–	–
Madagascar	16.0	40.4	29%	4.9%	47%	42%
Malawi	11.3	29.0	15%	4.6%	57%	76%
Malaysia	22.2	37.0	57%	2.9%	–	–
Mali	11.4	31.4	30%	5.1%	65%	69%
Malta	0.4	0.4	91%	0.7%	100%	100%
Mauritania	2.7	6.6	58%	5.1%	37%	33%
Mauritius	1.2	1.4	41%	1.6%	100%	99%
Mexico	98.9	146.6	74%	1.7%	88%	74%
Moldova	4.3	4.5	42%	0.0%	92%	99%
Mongolia	2.5	4.4	57%	1.3%	60%	30%
Morocco	29.9	45.4	55%	2.9%	80%	68%
Mozambique	18.3	42.9	32%	5.1%	57%	43%
Namibia	1.8	3.0	31%	3.3%	77%	41%
Nepal	23.0	49.3	12%	5.1%	88%	28%
Netherlands	15.9	14.2	89%	0.5%	100%	100%
New Zealand	3.8	5.2	86%	0.9%	–	–
Nicaragua	5.1	11.6	56%	3.3%	77%	85%
Niger	10.8	32.0	21%	6.0%	59%	20%
Nigeria	113.9	244.3	44%	4.4%	62%	54%
Norway	4.5	4.8	75%	0.7%	100%	–
Oman	2.5	8.3	76%	4.0%	39%	92%
Pakistan	141.3	345.5	33%	3.5%	90%	62%
Panama	2.9	4.3	56%	2.0%	90%	92%
Papua New Guinea	4.8	9.5	17%	3.7%	42%	82%
Paraguay	5.5	12.6	56%	3.6%	78%	94%
Peru	25.7	42.3	73%	2.1%	80%	71%
Philippines	75.7	130.9	59%	3.2%	86%	83%
Poland	38.6	36.3	62%	0.3%	–	–
Portugal	10.0	8.1	64%	1.9%	–	–
Qatar	0.6	0.8	93%	1.7%	–	–
Romania	22.4	16.4	55%	0.1%	58%	53%

Sources: **Col 1:** UN Population Division; **Col 2:** UN Millennium Indicator – UNICEF; UN Population Division; **Cols 3 & 4:** UNICEF;

NEEDS AND RESOURCES

5 Internal renewable water resources		6 Water dependency	7 Annual groundwater withdrawals	8 Desalination capacity	9 Water Poverty Index	Countries
m³ per person per year 2000	m³ per person per year 2050 projected	% of renewable water originating outside country 2000	m³ per person 2000	m³ per day 1996	rating out of 100 2003	
0	0	100%	143	1,284,327	54	Kuwait
9,439	6,162	0%	132	–	64	Kyrgyzstan
36,074	16,648	43%	–	–	54	Laos
6,916	9,599	53%	–	–	54	Latvia
1,373	957	1%	153	17,083	56	Lebanon
2,556	2,098	0%	–	–	43	Lesotho
68,656	13,918	14%	–	–	–	Liberia
113	60	0%	735	638,377	–	Libya
4,210	5,206	38%	55	–	–	Lithuania
2,289	1,399	68%	–	–	–	Luxembourg
2,950	3,168	16%	–	–	–	Macedonia
21,102	7,166	0%	483	–	48	Madagascar
1,548	562	7%	–	–	38	Malawi
26,104	15,324	0%	19	13,699	67	Malaysia
5,286	1,438	40%	12	–	41	Mali
40	39	0%	–	145,031	–	Malta
150	53	96%	498	4,440	50	Mauritania
1,894	1,543	0%	–	–	60	Mauritius
4,137	2,789	11%	275	105,146	58	Mexico
233	279	91%	–	–	49	Moldova
13,737	8,394	0%	149	–	55	Mongolia
1,004	596	0%	98	19,700	46	Morocco
5,467	2,575	54%	–	–	45	Mozambique
3,530	1,693	66%	–	1,090	60	Namibia
8,601	3,781	6%	–	–	54	Nepal
630	631	88%	70	110,438	69	Netherlands
86,554	73,665	0%	–	–	69	New Zealand
37,414	16,532	4%	–	–	58	Nicaragua
323	67	90%	18	–	35	Niger
1,941	793	23%	–	6,000	44	Nigeria
85,925	78,689	0%	98	1,200	77	Norway
388	113	0%	281	180,621	59	Oman
1,756	721	76%	490	4,560	58	Pakistan
51,623	34,589	0%	–	–	67	Panama
166,555	72,951	0%	–	–	55	Papua New Guinea
17,102	7,481	72%	–	1,000	56	Paraguay
62,973	38,365	16%	139	24,538	64	Peru
6,332	3,731	0%	83	5,648	61	Philippines
1,280	1,480	13%	52	20,564	56	Poland
3,794	4,219	45%	311	5,920	65	Portugal
90	61	4%	–	560,764	57	Qatar
1,649	2,039	80%	158	–	59	Romania

Cols 5, 6, & 7: FAO Aquastat 2003; Col 8: Pacific Institute <www.worldwater.org/table16.html>; Col 9: Peter Lawrence, Jeremy Meigh and Caroline Sullivan: *The Water Poverty Index: an International Comparison*, Keele Economic Research Papers 2002/19, March 2003

Countries	1 Total population millions		2 Urban population		3 Improved water source % of population with access 2000	4 Improved sanitation % of population with access 2000
	2000	2050 projected	as % of total population 2000	annual % increase 2000–2005 projected		
Russia	145.5	121.3	73%	–0.6%	99%	–
Rwanda	7.6	16.0	6%	4.2%	41%	8%
Samoa	0.4	0.4	22%	1.4%	99%	99%
Saudi Arabia	20.3	54.5	86%	3.6%	95%	100%
Senegal	9.4	23.1	47%	4.0%	78%	70%
Seychelles	0.1	0.1	64%	2.4%	–	–
Sierra Leone	4.4	11.0	37%	6.3%	57%	66%
Singapore	4.0	4.0	100%	1.7%	100%	100%
Slovakia	5.4	4.8	57%	0.4%	100%	100%
Slovenia	2.0	1.5	49%	–0.1%	100%	–
Solomon Islands	0.4	1.1	20%	6.0%	71%	34%
Somalia	8.8	31.8	27%	5.8%	–	–
South Africa	43.3	52.5	57%	2.1%	86%	87%
Spain	39.9	30.2	78%	0.3%	–	–
Sri Lanka	18.9	25.9	23%	2.4%	77%	94%
Sudan	31.1	59.2	36%	4.7%	75%	62%
Suriname	0.4	0.6	74%	1.3%	82%	93%
Swaziland	0.9	2.4	26%	2.2%	–	–
Sweden	8.8	8.7	83%	–0.1%	100%	100%
Switzerland	7.2	6.7	67%	0.0%	100%	100%
Syria	16.2	34.5	51%	3.3%	80%	90%
Tajikistan	6.1	11.3	28%	0.7%	60%	90%
Tanzania	35.1	80.6	32%	5.3%	68%	90%
Thailand	62.8	74.2	20%	2.1%	84%	96%
Togo	4.5	12.1	33%	4.2%	54%	34%
Trinidad & Tobago	1.3	1.5	74%	1.0%	90%	99%
Tunisia	9.5	15.0	66%	2.1%	80%	84%
Turkey	66.7	100.7	66%	1.9%	82%	90%
Turkmenistan	4.7	7.7	45%	2.3%	–	–
Uganda	23.3	64.9	14%	5.7%	52%	79%
Ukraine	49.6	39.3	68%	–0.8%	98%	99%
United Arab Emirates	2.6	3.6	87%	2.2%	–	–
United Kingdom	59.4	56.7	89%	0.3%	100%	100%
United States of America	283.2	349.3	77%	1.2%	100%	100%
Uruguay	3.3	4.4	92%	0.9%	98%	94%
Uzbekistan	24.9	40.6	37%	1.4%	85%	89%
Venezuela	24.2	42.2	87%	2.1%	83%	68%
Vietnam	78.1	126.8	24%	3.1%	77%	47%
Yemen	18.3	58.8	25%	5.3%	69%	38%
Yugoslavia	10.6	10.5	52%	0.2%	98%	100%
Zambia	10.4	21.2	40%	2.7%	64%	78%
Zimbabwe	12.6	18.1	35%	3.7%	83%	62%

Sources: Col 1: UN Population Division; **Col 2:** UN Millennium Indicator – UNICEF; UN Population Division; **Cols 3 & 4:** UNICEF;

NEEDS AND RESOURCES

5 Internal renewable water resources		6 Water dependency	7 Annual groundwater withdrawals	8 Desalination capacity	9 Water Poverty Index	Countries
m³ per person per year 2000	m³ per person per year 2050 projected	% of renewable water originating outside country 2000	m³ per person 2000	m³ per day 1996	rating out of 100 2003	
29,642	41,366	4%	86	116,140	63	Russia
828	340	0%	–	–	39	Rwanda
	–	0%	–	–	–	Samoa
118	40	0%	899	5,006,194	53	Saudi Arabia
2,802	1,162	33%	39	–	45	Senegal
	–	0%	–	–	–	Seychelles
36,325	11,149	0%	–	–	42	Sierra Leone
149	130	0%	–	133,695	56	Singapore
5,703	6,588	75%	113	–	71	Slovakia
9,307	12,115	41%	89	–	69	Slovenia
99,904	30,658	0%	–	–	–	Solomon Islands
684	147	56%	46	–	–	Somalia
1,034	947	10%	65	79,531	52	South Africa
2,764	3,526	0%	137	492,824	64	Spain
2,642	2,168	0%	–	–	56	Sri Lanka
1,126	551	77%	13	1,450	49	Sudan
210,951	210,526	28%	–	–	75	Suriname
2,811	1,869	41%	–	–	53	Swaziland
19,905	22,631	2%	73	1,300	72	Sweden
5,927	7,580	24%	126	2,506	72	Switzerland
432	193	80%	134	5,488	55	Syria
10,892	6,791	17%	399	–	59	Tajikistan
2,278	967	10%	–	–	48	Tanzania
3,344	2,546	49%	15	24,075	64	Thailand
2,540	972	22%	–	–	46	Togo
2,967	2,787	0%	–	–	59	Trinidad & Tobago
372	250	9%	182	47,402	51	Tunisia
2,940	1,983	1%	124	600	57	Turkey
287	162	97%	100	43,707	70	Turkmenistan
1,674	384	41%	–	–	44	Uganda
1,071	1,772	62%	78	21,000	–	Ukraine
58	40	0%	724	2,134,233	52	United Arab Emirates
2,440	2,460	1%	42	101,397	72	United Kingdom
8,682	6,193	8%	432	2,799,000	65	United States of America
17,680	13,886	58%		–	67	Uruguay
657	403	77%	334	31,200	61	Uzbekistan
29,891	17,139	41%	–	19,629	65	Venezuela
4,690	2,961	59%	12	–	52	Vietnam
223	40	0%	139	36,996	44	Yemen
4,170	4,873	79%		2,204	–	Yugoslavia
7,696	2,741	24%	–	–	50	Zambia
1,117	599	30%	–	–	53	Zimbabwe

Cols 5, 6, & 7: FAO Aquastat 2003; Col 8: Pacific Institute <www.worldwater.org/table16.html>; Col 9: Peter Lawrence, Jeremy Meigh and Caroline Sullivan: *The Water Poverty Index: an International Comparison*, Keele Economic Research Papers 2002/19, March 2003

Countries	1 Water use m³ per person 2000			2 Water use per sector as % of total use 2000			3 Irrigation as % of total arable and permanent crop area 2000
	Domestic	Agriculture	Industry	Domestic	Agriculture	Industry	
Afghanistan	19	1,050	0	2%	98%	0%	30%
Albania	146	340	61	27%	62%	11%	49%
Algeria	44	130	26	22%	65%	13%	7%
Angola	6	16	4	22%	61%	16%	2%
Argentina	129	581	74	16%	74%	9%	6%
Armenia	234	512	34	30%	66%	4%	51%
Australia	184	941	125	15%	75%	10%	5%
Austria	92	3	167	35%	1%	64%	0%
Azerbaijan	103	1,448	593	5%	68%	28%	76%
Bahamas	–	–	–	–	–	–	–
Bahrain	185	266	17	40%	57%	4%	53%
Bangladesh	18	555	4	3%	96%	1%	44%
Belarus	64	82	127	23%	30%	46%	2%
Belgium	–	–	719	13%*	1%*	85%*	5%*
Belize	62	1	489	11%	0%	89%	3%
Benin	6	30	4	15%	74%	11%	0%
Bhutan	7	192	2	4%	95%	1%	24%
Bolivia	22	139	6	13%	83%	3%	6%
Bosnia-Herzegovina	–	–	–	–	–	–	0%
Botswana	35	39	17	38%	43%	19%	0%
Brazil	71	215	62	20%	62%	18%	4%
Brunei	–	–	–	–	–	–	14%
Bulgaria	40	248	1,033	3%	19%	78%	17%
Burkina Faso	8	60	0	11%	88%	0%	1%
Burma	9	684	4	1%	98%	1%	15%
Burundi	6	30	0	17%	82%	1%	1%
Cambodia	5	306	2	2%	98%	1%	7%
Cameroon	12	49	5	18%	74%	8%	0%
Canada	292	176	1,026	20%	12%	69%	2%
Central African Rep.	5	0	1	77%	4%	19%	0%
Chad	6	24	0	19%	80%	1%	0%
Chile	93	524	208	11%	64%	25%	83%
China	32	333	126	7%	68%	26%	39%
Colombia	128	117	10	50%	46%	4%	20%
Comoros	–	–	–	–	–	–	0%
Congo	8	1	4	59%	10%	30%	0%
Congo, Dem. Rep.	4	2	1	52%	31%	16%	0%
Costa Rica	196	356	114	29%	53%	17%	20%
Côte d'Ivoire	14	38	7	23%	65%	12%	1%
Croatia	–	–	–	–	–	–	0%
Cuba	139	504	89	19%	69%	12%	19%
Cyprus	85	222	4	27%	71%	1%	28%

Sources: Columns 1 – 3: FAO Aquastats 2003;   * The percentage given for Belgium and Luxembourg is a composite figure for the two countries together

4 Hydropower as % of total power *2000*	5 Phosphorous concentration in water mg/litre *1996 or latest available estimated data in italics*	6 Organic water pollutants emitted kg per worker per day *2000*	7 Wetlands of International Importance total area in thousand hectares *2002*	Countries
–	–	–	–	Afghanistan
99%	0	29	20	Albania
0%	0.4	24	1,866	Algeria
63%	0.57	20	–	Angola
32%	0.04	21	2,670	Argentina
21%	0.48	25	492	Armenia
8%	0.06	21	5,310	Australia
70%	0.1	13	118	Austria
8%	0.6	17	100	Azerbaijan
–	–	–	–	Bahamas
–	–	–	–	Bahrain
6%	0.51	14	606	Bangladesh
0%	0.36	–	204	Belarus
1%	1.63	16	8	Belgium
–	–	–	7	Belize
–	0.67	–	139	Benin
–	0.13	–	–	Bhutan
50%	0.34	25	5,504	Bolivia
49%	0.36	18	7	Bosnia-Herzegovina
–	0.2	20	6,864	Botswana
87%	0.09	20	6,346	Brazil
–	–	–	–	Brunei
7%	0.39	17	3	Bulgaria
–	0.38	22	299	Burkina Faso
37%	0.31	13	–	Burma
–	0.68	24	1	Burundi
–	0.43	16	55	Cambodia
99%	0.5	20	–	Cameroon
59%	0	15	13,052	Canada
–	0.35	17	–	Central African Rep.
–	0.36	–	1,843	Chad
46%	0.51	24	100	Chile
16%	0.28	14	2,548	China
73%	0.36	21	439	Colombia
–	–	–	–	Comoros
100%	–	–	439	Congo
100%	0.21	–	866	Congo, Dem. Rep.
82%	0.34	22	313	Costa Rica
37%	0.14	24	19	Côte d'Ivoire
55%	0.5	17	81	Croatia
1%	0.01	–	452	Cuba
–	–	–	–	Cyprus

Col 4: World Bank *World Development Indicators 2003*, Indicator 3.9; **Col 5:** <www.nationmaster.com>; **Col 6:** World Bank *World Development Indicators 2003*, Indicator 3.6; **Col 7:** EarthTrends Data Tables: Biodiversity and Protected Areas 2003 on <http://earthtrends.wri.org>

Countries	1 Water use m³ per person 2000			2 Water use per sector as % of total use 2000			3 Irrigation as % of total arable and permanent crop area 2000
	Domestic	Agriculture	Industry	Domestic	Agriculture	Industry	
Czech Republic	102	5	143	41%	2%	57%	1%
Denmark	77	101	61	32%	42%	26%	21%
Djibouti	1	11	0	11%	89%	0%	67%
Dominican Republic	130	268	7	32%	66%	2%	17%
Ecuador	167	1,104	71	12%	82%	5%	29%
Egypt	77	793	141	8%	78%	14%	99%
El Salvador	50	121	32	25%	59%	16%	6%
Equatorial Guinea	192	2	37	83%	1%	16%	–
Eritrea	4	79	1	4%	95%	1%	6%
Estonia	65	6	46	56%	5%	39%	0%
Ethiopia	0	39	2	1%	93%	6%	2%
Fiji	9	66	9	11%	78%	11%	1%
Finland	65	13	401	14%	3%	84%	3%
France	106	66	502	16%	10%	74%	10%
Gabon	50	42	12	48%	40%	11%	1%
Gambia	5	16	3	22%	67%	11%	1%
Georgia	137	405	144	20%	59%	21%	44%
Germany	71	114	389	12%	20%	68%	4%
Ghana	10	13	4	37%	48%	15%	0%
Greece	119	589	23	16%	81%	3%	37%
Guatemala	11	141	24	6%	80%	13%	7%
Guinea	14	167	4	8%	90%	2%	6%
Guinea-Bissau	8	83	1	9%	91%	1%	5%
Guyana	36	2,102	19	2%	97%	1%	30%
Haiti	6	114	1	5%	94%	1%	10%
Honduras	11	108	15	8%	81%	11%	5%
Hungary	71	246	450	9%	32%	59%	4%
Iceland	187	1	361	34%	0%	66%	–
India	52	553	35	8%	86%	5%	30%
Indonesia	31	356	3	8%	91%	1%	13%
Iran	70	942	24	7%	91%	2%	44%
Iraq	59	1,716	86	3%	92%	5%	64%
Ireland	67	0	230	23%	0%	77%	–
Israel	104	211	23	31%	63%	7%	43%
Italy	140	348	283	18%	45%	37%	25%
Jamaica	54	78	27	34%	49%	17%	9%
Japan	137	435	124	20%	62%	18%	65%
Jordan	43	155	9	21%	75%	4%	16%
Kazakhstan	37	1,770	358	2%	82%	17%	16%
Kenya	15	33	3	30%	64%	6%	1%
Korea (North)	81	223	102	20%	55%	25%	73%
Korea (South)	142	191	65	36%	48%	16%	46%

Sources: Columns 1 – 3: FAO Aquastats 2003;   * The percentage given for Belgium and Luxembourg is a composite figure for the two countries together

USES AND ABUSES

4 Hydropower as % of total power 2000	5 Phosphorous concentration in water mg/litre 1996 or latest available estimated data in italics	6 Organic water pollutants emitted kg per worker per day 2000	7 Wetlands of International Importance total area in thousand hectares 2002	Countries
2%	0.29	14	42	Czech Republic
0%	0.14	17	2,283	Denmark
8%	–	–	–	Djibouti
72%	–	–	20	Dominican Republic
19%	0.25	27	83	Ecuador
31%	0.6	19	106	Egypt
–	0.22	18	2	El Salvador
–	–	–	–	Equatorial Guinea
–	–	–	–	Eritrea
0%	0.11	–	216	Estonia
97%	0.38	–	–	Ethiopia
–	–	–	–	Fiji
21%	0.01	19	139	Finland
12%	0.17	10	795.1	France
71%	0.29	26	1,080	Gabon
–	0.53	34	20	Gambia
79%	–	–	34	Georgia
4%	0.32	13	829	Germany
92%	0.13	17	178	Ghana
7%	0.31	20	164	Greece
38%	0.41	28	503	Guatemala
–	0.49	–	4,779	Guinea
–	0.82	–	39	Guinea-Bissau
–	–	–	–	Guyana
52%	0.34	–	–	Haiti
62%	0.4	20	172	Honduras
1%	0.21	17	154	Hungary
–	0.35	–	59	Iceland
14%	0.15	19	195	India
10%	0.56	18	243	Indonesia
3%	0.35	17	1,476	Iran
2%	0.01	16	–	Iraq
4%	0.11	15	67	Ireland
0%	0.42	16	0	Israel
16%	0.13	13	57	Italy
2%	1.01	29	6	Jamaica
8%	0.06	15	84	Japan
1%	1.01	18	7	Jordan
15%	0.47	–	–	Kazakhstan
34%	0.58	25	91	Kenya
67%	0.81	–	–	Korea (North)
1%	1.13	12	1	Korea (South)

Col 4: World Bank *World Development Indicators 2003*, Indicator 3.9; Col 5: <www.nationmaster.com>; Col 6: World Bank *World Development Indicators 2003*, Indicator 3.6; Col 7: EarthTrends Data Tables: Biodiversity and Protected Areas 2003 on <http://earthtrends.wri.org>

Countries	1 Water use m³ per person 2000			2 Water use per sector as % of total use 2000			3 Irrigation as % of total arable and permanent crop area 2000
	Domestic	Agriculture	Industry	Domestic	Agriculture	Industry	
Kuwait	104	122	7	45%	52%	3%	48%
Kyrgyzstan	64	1,921	63	3%	94%	3%	75%
Laos	24	511	32	4%	90%	6%	16%
Latvia	66	15	40	55%	12%	33%	1%
Lebanon	128	262	2	33%	67%	1%	26%
Lesotho	11	5	11	40%	19%	41%	1%
Liberia	10	21	6	28%	56%	15%	0%
Libya	75	807	27	8%	89%	3%	22%
Lithuania	56	5	11	78%	7%	15%	0%
Luxembourg	–	–	–	13%*	1%*	85%*	5%*
Macedonia	–	–	–	–	–	–	9%
Madagascar	26	896	15	3%	96%	2%	31%
Malawi	13	72	4	15%	81%	5%	1%
Malaysia	68	252	86	17%	62%	21%	5%
Mali	4	605	1	1%	99%	0%	2%
Malta	105	36	1	74%	25%	1%	8%
Mauritania	56	563	18	9%	88%	3%	10%
Mauritius	134	318	75	25%	60%	14%	17%
Mexico	137	610	43	17%	77%	5%	23%
Moldova	51	178	309	9%	33%	58%	14%
Mongolia	36	91	49	20%	52%	28%	7%
Morocco	36	384	7	8%	90%	2%	13%
Mozambique	4	30	1	11%	87%	2%	3%
Namibia	50	96	7	33%	63%	5%	1%
Nepal	13	426	3	3%	96%	1%	38%
Netherlands	31	170	300	6%	34%	60%	60%
New Zealand	271	235	53	49%	42%	9%	9%
Nicaragua	37	213	7	14%	83%	3%	2%
Niger	9	192	1	4%	95%	1%	1%
Nigeria	15	48	7	21%	69%	10%	1%
Norway	111	51	327	23%	10%	67%	14%
Oman	38	484	11	7%	91%	2%	77%
Pakistan	23	1,151	25	2%	96%	2%	72%
Panama	192	82	15	66%	28%	5%	5%
Papua New Guinea	9	0	7	56%	1%	43%	–
Paraguay	18	64	8	20%	72%	9%	3%
Peru	66	640	79	8%	82%	10%	28%
Philippines	63	279	36	17%	74%	9%	15%
Poland	54	35	330	13%	8%	79%	1%
Portugal	108	880	136	10%	78%	12%	23%
Qatar	132	373	15	25%	72%	3%	60%
Romania	89	589	355	9%	57%	34%	29%

Sources: Columns 1 – 3: FAO Aquastats 2003;   * The percentage given for Belgium and Luxembourg is a composite figure for the two countries together

USES AND ABUSES

4 Hydropower as % of total power 2000	5 Phosphorous concentration in water mg/litre 1996 or latest available estimated data in italics	6 Organic water pollutants emitted kg per worker per day 2000	7 Wetlands of International Importance total area in thousand hectares 2002	Countries
–	0.66	17	–	Kuwait
92%	0.23	16	–	Kyrgyzstan
–	0.45	–	–	Laos
68%	0.1	21	43	Latvia
6%	0.38	19	1	Lebanon
–	–	16	–	Lesotho
–	0.49	–	–	Liberia
–	0.47	–	–	Libya
3%	0.08	18	51	Lithuania
–	–	–	–	Luxembourg
–	0.34	18	19	Macedonia
–	0.45	–	53	Madagascar
–	0.52	29	225	Malawi
10%	0.04	11	38	Malaysia
–	0.15	–	162	Mali
–	–	–	–	Malta
–	0.48	–	1,231	Mauritania
–	–	15	–	Mauritius
16%	0.64	20	1,157	Mexico
2%	0.2	29	19	Moldova
–	0.17	18	631	Mongolia
5%	0.26	17	14	Morocco
100%	0.49	16	–	Mozambique
98%	0.35	35	630	Namibia
98%	0.42	14	18	Nepal
0%	0.27	18	327	Netherlands
63%	0.04	22	39	New Zealand
9%	0.61	–	406	Nicaragua
–	0.69	–	715	Niger
37%	0.66	17	58	Nigeria
99%	0.01	20	70	Norway
–	0.2	17	–	Oman
25%	0.2	18	284	Pakistan
67%	0.37	31	111	Panama
–	0.11	–	595	Papua New Guinea
100%	0.18	28	775	Paraguay
81%	0.29	21	6,759	Peru
17%	0.35	18	68	Philippines
1%	0.33	16	91	Poland
26%	0.13	14	66	Portugal
–	–	–	–	Qatar
28%	0.4	14	665	Romania

Col 4: World Bank *World Development Indicators 2003*, Indicator 3.9; Col 5: <www.nationmaster.com>; Col 6: World Bank *World Development Indicators 2003*, Indicator 3.6; Col 7: EarthTrends Data Tables: Biodiversity and Protected Areas 2003 on <http://earthtrends.wri.org>

Countries	1 Water use m³ per person 2000			2 Water use per sector as % of total use 2000			3 Irrigation as % of total arable and permanent crop area 2000
	Domestic	Agriculture	Industry	Domestic	Agriculture	Industry	
Russia	99	94	334	19%	18%	63%	5%
Rwanda	5	4	1	–	–	–	0%
Samoa	–	–	–	–	–	–	–
Saudi Arabia	83	758	10	10%	89%	1%	42%
Senegal	10	152	6	6%	90%	4%	3%
Seychelles	–	–	–	–	–	–	–
Sierra Leone	4	80	2	5%	93%	2%	5%
Singapore	–	–	–	–	–	–	–
Slovakia	–	–	–	–	–	–	11%
Slovenia	–	–	–	–	–	–	1%
Solomon Islands	–	–	–	–	–	–	–
Somalia	2	374	0	0%	100%	0%	19%
South Africa	59	257	37	17%	73%	10%	8%
Spain	120	607	165	13%	68%	19%	20%
Sri Lanka	16	634	16	2%	95%	2%	30%
Sudan	32	1,160	8	3%	97%	1%	12%
Suriname	72	1,477	46	4%	93%	3%	76%
Swaziland	25	822	51	3%	92%	6%	35%
Sweden	123	30	182	37%	9%	54%	4%
Switzerland	87	7	265	24%	2%	74%	6%
Syria	41	1,169	22	3%	95%	2%	19%
Tajikistan	73	1,801	92	4%	92%	5%	84%
Tanzania	3	53	1	6%	93%	1%	3%
Thailand	35	1,318	34	2%	95%	2%	28%
Togo	17	17	3	–	–	–	0%
Trinidad & Tobago	159	13	64	67%	6%	27%	3%
Tunisia	45	236	7	16%	82%	2%	8%
Turkey	83	418	62	15%	74%	11%	16%
Turkmenistan	88	5,075	40	2%	98%	1%	97%
Uganda	6	5	2	45%	39%	15%	0%
Ukraine	92	397	268	12%	52%	35%	8%
United Arab Emirates	204	604	77	23%	68%	9%	27%
United Kingdom	35	5	121	22%	3%	75%	2%
United States of America	215	698	779	13%	41%	46%	12%
Uruguay	23	909	11	2%	96%	1%	14%
Uzbekistan	111	2,185	48	5%	93%	2%	88%
Venezuela	157	164	24	45%	47%	7%	17%
Vietnam	71	622	221	–	–	–	41%
Yemen	15	344	2	4%	95%	1%	29%
Yugoslavia	–	–	–	–	–	–	2%
Zambia	27	127	13	16%	76%	8%	1%
Zimbabwe	20	178	10	10%	86%	5%	3%

Sources: Columns 1 – 3: FAO Aquastats 2003; * The percentage given for Belgium and Luxembourg is a composite figure for the two countries together

USES AND ABUSES

4 Hydropower as % of total power 2000	5 Phosphorous concentration in water mg/litre 1996 or latest available estimated data in italics	6 Organic water pollutants emitted kg per worker per day 2000	7 Wetlands of International Importance total area in thousand hectares 2002	Countries
19%	0.14	16	10,324	Russia
–	0.49	–	–	Rwanda
–	–	–	–	Samoa
–	0.11	14	–	Saudi Arabia
–	0.34	30	100	Senegal
–	–	–	–	Seychelles
–	0.36	32	295	Sierra Leone
–	–	9	–	Singapore
16%	0.22	15	38	Slovakia
28%	0.1	17	1	Slovenia
–	–	–	–	Solomon Islands
–	0.35	–	–	Somalia
1%	0.73	17	499	South Africa
13%	0.5	15	158	Spain
47%	0.1	18	8	Sri Lanka
48%	1.75	–	–	Sudan
–	–	–	12	Suriname
–	–	23	–	Swaziland
54%	0.28	14	515	Sweden
56%	0.07	17	7	Switzerland
41%	0.21	20	10	Syria
98%	0.96	–	95	Tajikistan
96%	0.32	25	4,272	Tanzania
6%	0.31	16	132	Thailand
2%	0.33	–	194	Togo
–	0.08	28	6	Trinidad & Tobago
1%	0.39	16	13	Tunisia
25%	0.35	17	159	Turkey
0%	0.48	–	–	Turkmenistan
–	0.16	–	15	Uganda
7%	0.23	18	716	Ukraine
–	0.44	–	–	United Arab Emirates
1%	0.09	15	855	United Kingdom
6%	0.08	12	1,190	United States of America
93%	0.31	27	407	Uruguay
13%	0.51	–	–	Uzbekistan
74%	0.45	21	264	Venezuela
55%	0.59	–	12	Vietnam
–	–	25	–	Yemen
38%	–	16	40	Yugoslavia
99%	0.56	22	333	Zambia
47%	0.09	20	–	Zimbabwe

Col 4: World Bank *World Development Indicators 2003*, Indicator 3.9; Col 5: <www.nationmaster.com>; Col 6: World Bank *World Development Indicators 2003*, Indicator 3.6; Col 7: EarthTrends Data Tables: Biodiversity and Protected Areas 2003 on <http://earthtrends.wri.org>

GLOSSARY

access to an improved source – an **improved water source** providing at least 20 litres per person per day from a source within 1 kilometre of the dwelling

aquifer – a natural underground layer, often of sand or gravel, that contains water

billion – a thousand million

brackish water – water that is neither **fresh** nor **salt**

desalination – the changing of **salt** or **brackish** water into **fresh** water; *see also* **distillation** and **reverse osmosis**

distillation – method of **desalination** in which water is boiled to steam and condensed in a separate reservoir, leaving behind contaminants with higher boiling points than water

evaporation – the process of liquid water becoming water vapour, including vaporization from water surfaces, land surfaces, and snow fields, but not from leaf surfaces

evapotranspiration – both **evaporation** and **transpiration** (the process by which water is evaporated from a plant surface, such as leaf pores)

fresh water – water that contains fewer than 1,000 milligrams per litre of dissolved solids

groundwater – water that is pumped from **aquifers**

improved sanitation – adequate facilities for the disposal of human excreta, including a connection to a **sewer** or **septic tank** system, a **pour-flush latrine**, a simple **pit latrine** or a **ventilated improved pit latrine**; it is considered adequate if it is private or shared (but not public) and if it can effectively prevent human, animal and insect contact with faeces

improved water source – a household connection, public standpipe, borehole, protected well or spring, or rainwater collection

internal renewable water resources – average annual flow of rivers and recharge of groundwater generated from **precipitation** falling within the country's borders

leaching – the process by which soluble materials in the soil, such as salts, nutrients, pesticide chemicals or contaminants, are washed into a lower layer of soil, or are dissolved and carried away by water.

levee – a natural or artificial earthen barrier along the edge of a stream, lake, or river

pit latrine – a pit used as a toilet

pour-flush latrine – a type of latrine or toilet in which water is poured in to flush away human waste; some of the water remains in a U-shaped pipe to prevent flies and mosquitoes from making contact with the human faeces

precipitation – rain, snow, hail, sleet, dew, and frost

renewable resources – total resources offered by the average annual natural inflow and runoff that feed a catchment area or aquifer; natural resources that, after exploitation, can return to their previous stock levels by the natural processes of growth or replenishment

reverse osmosis – a desalination process that uses a semi-permeable membrane to separate and remove dissolved solids, viruses, bacteria and other matter from water; **salt** or **brackish** water is forced across a membrane, leaving the impurities behind and creating **fresh water**

salt water – water that contains significant amounts of dissolved solids

sewer – a pipe used to carry off waste matter

surface water – water pumped from sources open to the atmosphere, such as rivers, lakes, and reservoirs

unimproved water source – vendor, tanker trucks, and unprotected wells and springs

septic tank – a tank in which sewage is deposited and retained until it has been disintegrated by bacteria

ventilated improved pit latrine – a dry latrine system, with a screened vent pipe to trap flies, and often with double pits to allow use on a permanent rotating basis

wastewater treatment – *primary treatment*: the first stage of the wastewater-treatment process, in which filters and scrapers are used to remove pollutants; *secondary treatment*: the removal and further reduction of contaminants and effluent, including about 90 percent of the oxygen-demanding substances and suspended solids; disinfection is the final stage of secondary treatment; *tertiary treatment*: biological, physical, and chemical separation processes to remove substances that have resisted the previous two treatments

water table – the upper level of groundwater in soil

withdrawal – water removed from groundwater or surface water for use

USEFUL CONVERSIONS

1 cubic metre (m³) = 1,000 litres

1 cubic kilometre (km³) = 1,000,000,000 cubic metres (m³) = 1,000,000,000,000 litres

1 litre = 0.264 US gallons (liquid) = 0.219 UK gallons

1 US gallon (liquid) = 3.785 litres

1 UK gallon = 4.55 litres

1 cubic metre (m³) = 264.172 US gallons (liquid) = 219.9 UK gallons

1 US gallon (liquid) = 0.00378 cubic metres = 3,785 cubic centimetres (cc)

1 UK gallon = 0.00454 cubic metres = 4,546 cubic centimetres (cc)

1 cubic kilometre (km³) = 810,713 acre feet

1 acre foot = 1,233 cubic metres (m³) = 325,851 US gallons (liquid)

1 kilometre (km) = 0.621 miles

1 mile = 1.6 kilometres (km)

1 kilogram (kg) = 2.2 pounds (lb)

1 pound (lb) = 0.45 kilograms (kg) = 450 grams (g)

Metric water–weight conversion

1 kilogram (kg) of water = 1 litre of water

1 gram of water = 1 cubic centimetre (cc) of water

1 metric tonne (mt) of water = 1,000 kilogram (kg) of water = 1,000 litres of water = 1 cubic metre (m³)

USEFUL SOURCES

Books and papers

Janet Abramovitz 'Unnatural Disasters'. WorldWatch Paper 158, October 2001, p17

Maude Barlow and Tony Clarke *Blue Gold*. New Press, New York and Earthscan, London 2002

Meredith A. Giordano and Aaron T. Wolf 'The World's International Freshwater Agreements: Historical Developments and Future Opportunities' *Atlas of International Freshwater Agreements*. UNEP and FAO, 2002 <www.transboundarywaters.orst.edu/publications/atlas/>

Peter H. Gleick et al *The World's Water 2002–2003*. Island Press, 2002

Payal Sampat, 'Deep Trouble: The Hidden Threat of Groundwater Pollution'. World Watch Paper 154, December 2000

World Development Indicators 2003. World Bank, Washington DC 2003

The World Health Report 2002. WHO, Geneva 2003

World Water Assessment Programme *Water for People: Water for Life*. UNESCO and Berghahn Books 2003

World Water Vision. Earthscan, London 2000

Websites

An extensive list of useful websites is available on <www.worldwater.org/links.htm>. Those listed below were found especially useful in the preparation of this book.

Earth Policy Institute <www.earth-policy.org>

Earthtrends: The environmental information portal of the World Resources Insitute <earthtrends.wri.org>

Food and Agriculture Organisation (FAO): Aquastat 2003 <www.fao.org/ag/agl/aglw/aquastat/main/>

International Federation of the Red Cross, World Disaster Report 2002, Chaper 8 statistical analysis 1999–2002 <www.cred.be/emdat/intro.htm>

International Rivers Network < www.irn.org/>

International Water Management Institute <www.iwmi.cgiar.org>

International Water and Sanitation Centre <www.irc.nl/>

Nation Master <www.nationmaster.com>

Pacific Institute: Studies in Development, Environment and Security, Director Peter H. Gleick <www.pacinst.org/>

Population Report: Solutions for a Water-Short World. John Hopkins School for Public Health <www.infoforhealth.org/pr/m14edsum.shtml>

ReliefWeb <www.reliefweb.int>

Transboundary Freshwater Dispute Database <www.transboundarywaters.orst.edu/>

United Nations Environment Programme (UNEP): Vital Water Graphics <www.unep.org/vitalwater>

United Nations Environment Programme (UNEP) GRID Arendal, Norway <www.grida.no>

WaterAid <www.wateraid.org.uk/>

World Commission on Dams <www.dams.org>

Water Conflict Chronology, Peter H Gleick, Pacific Institute <www.worldwater.org/conflict.htm>

World Health Organisation (WHO) information on water-related disease: <www.who.int/water_sanitation_health/diseases/en>

World Water Assessment Programme Facts and Figures <www.unesco.org/water/wwap/facts_figures/index.shtml>

World Water Resources and their Use <webworld.unesco.org/water/ihp/db/shiklomanov/index.shtml> prepared by Professor Igor A Shiklomanov of the State Hydrological Institute (SHI)

Water Resources eAtlas: *Watersheds of the World*. <www.iucn.org/themes/wani/eatlas/>

World Water Council <www.worldwatercouncil.org/>

REFERENCES

Part 1: A Finite Resource

20–21 FRESH OUT OF WATER
The world's water
Freshwater sources
I. A. Shiklomanov 'World fresh water resources'
in Peter H. Gleick (ed.) *Water in Crisis: A guide
to the world's fresh water resources*. Oxford
University Press, New York, 1993

<webworld.unesco.org/water/ihp/db/shiklomanov
/summary/html/figure_2.html>

Peter H. Gleick *The World's Water 2000-2001*.
Island Press, 2000

22–23 MORE PEOPLE, LESS WATER
Water shortage
Food and Agriculture Organisation (FAO):
Aquastat 2003
<www.fao.org/ag/agl/aglw/aquastat/main/>

United Nations Population Division *World
Population Prospects 1950–2050: The 2000
Revision*. United Nations, New York, 2001

Running dry in the USA
Population Action International
<www.cnie.org/pop/pai/water-31.html>

Have and have nots
FAO Aquastat 2003
<www.fao.org/ag/agl/aglw/aquastat/main/>

United Nations Population Division *World
Population Prospects 1950–2050: The 2000
Revision*. United Nations, New York, 2001

Future water shortage
United Nations Population Division *World
Population Prospects 1950–2050: The 2000
Revision*. United Nations, New York, 2001

Food and Agriculture Organisation (FAO):
Aquastat 2003

<www.fao.org/ag/agl/aglw/aquastat/main/>

24–25 RISING DEMAND
What water is used for
World water use
FAO: Aquastat 2003
<www.fao.org/ag/agl/aglw/aquastat/main/>

Increasing use
<webworld.unesco.org/water/ihp/db/shiklomanov
/index.shtml>

26–27 ROBBING THE BANK
Payal Sampat 'Groundwater Shock', *World
Watch*, Jan/Feb 2000

World Water Assessment Programme *Water for
People, Water for Life*, UNESCO and Berghahn
Books 2003 pp 78–81

World Global Trends <www.earth-policy.org>

Groundwater withdrawals
World Resources 2000-2001, World Resources
Institute, 2000, Table FW.2. Available as
Groundwater and Desalinization 2000
<earthtrends.wri.org/datatables/index.cfm?theme
=2&CFID=427772&CFTOKEN=67640017>

Sandra Postel 'When the World's Wells Run Dry'
World Watch Sept/Oct 1999

Groundwater for drinking
Payal Sampat 'Groundwater Shock', *World
Watch*. Jan/Feb 2000, p12

Groundwater for irrigation
Burke and Moench, 2000; Foster et al, 2000,
cited in World Water Assessment Programme,
Water for People, Water for Life. UNESCO and
Berghahn Books 2003 p80

Part 2: Uses and Abuses

30–31 WATER AT HOME
Domestic water use
FAO Aquastat 2003
<www.fao.org/ag/agl/aglw/aquastat/main/>

Increasing domestic use
I. A. Shiklomanov <webworld.unesco.org/water/ihp/db/shiklomanov/index.shtml>

Household water consumption
Environment Canada
<www.ec.gc.ca/water/en/e_quickfacts.htm>

32–33 WATER FOR FOOD
Agricultural water use
FAO: Aquastat 2003
<www.fao.org/ag/agl/aglw/aquastat/main/>

<www.fao.org/docrep/005/y73523/y7352e00.htm>

Water for food
World Water Assessment Programme *Water for People, Water for Life*. UNESCO and Berghahn Books 2003

Peter H. Gleick *The World's Water Resources 2000–2001*. Island Press 2000

34–35 IRRIGATION
Irrigated land
Land turned salty by irrigation
FAO: Aquastat 2003
<www.fao.org/ag/agl/aglw/aquastat/main/>

Increasing irrigation
GEO Data Portal

Decline of the Aral Sea
UNEP: Vital Water Graphics
<www.unep.org/vitalwater/25.htm>

Leakage
<waterconservation.ifas.ufl.edu/trivia.htm>

Salinity
<www.fao.org/docrep/U8480E/U8480E0c.htm>

36–37 AGRICULTURAL POLLUTION
Payal Sampat, 'Groundwater Shock', *World Watch*. January/February 2000

Payal Sampat, 'Deep Trouble: The Hidden Threat of Groundwater Pollution'. WorldWatch Paper 154, December 2000

Fertilizers in water
UNEP Global Environmental Monitoring System/Water Quality monitoring System <www.cciw.ca/gems/> with data for additional 29 countries from R Prescott-Allen. Data, downloaded from <www.nationmaster.com>

Payal Sampat, 'Groundwater Shock', *World Watch*. January/February 2000

<wa.water.usgs.gov/news/2000/news.ofr99-244.htm>

Trends in fertilizer use
International Fertilizer Industry Association <www.fertilizer.org>

Animal slurry
Payal Sampat, 'Groundwater Shock', *World Watch*. January/February 2000

38–39 WATER FOR INDUSTRY
Industrial water use
Heavy industrial users
FAO: Aquastat 2003
<www.fao.org/ag/agl/aglw/aquastat/main/>

Increasing industrial use
<webworld.unesco.org/water/ihp/db/shiklomanov/index.shtml>

Making water work
World Water Assessment Programme Facts and Figures <www.unesco.org/water/wwap/facts_figures/index.shtml>

Adding value
Payal Sampat, 'Groundwater Shock', *World Watch*. January/February 2000

40–41 INDUSTRIAL POLLUTION
Payal Sampat, 'Groundwater Shock', *World Watch*. January/February 2000

Payal Sampat, 'Deep Trouble: The Hidden Threat of Groundwater Pollution'. WorldWatch Paper 154, December 2000

Organic pollutants
World Development Indicators 2003. World Bank, Washington DC 2003, Indicator 3.6

Major polluters
World Development Indicators 2003. World Bank, Washington DC 2003, Indicator 3.6

Polluting industries
World Development Indicators 2001, cited in World Water Assessment Programme *Water for People, Water for Life*. UNESCO and Berghahn Books, 2003. p 299

Contaminated aquifers
Payal Sampat, 'Groundwater Shock', *World Watch*. January/February 2000

Payal Sampat, 'Deep Trouble: The Hidden Threat of Groundwater Pollution'. WorldWatch Paper 154, December 2000

Industrial sludge
<www.unesco.org/water/wwap/facts_figures/water_industry.shtml>

Industrial India
quoted by Payal Sampat, from Manish Tiwari and Richard Mahapatra, 'What Goes Down Must Come Up', *Down to Earth*. 31 August 1999, pp30–40 <www.downtoearth.org.in>

Yangtze River
Maude Barlow and Tony Clarke *Blue Gold*. New Press, New York and Earthscan, London 2002

42–43 WATER FOR POWER
Importance of hydropower
Percentage change in share
World Development Indicators 2003. World Bank, Washington DC 2003, Indicator 3.9

Hydropower in China
<www.choose-positive-energy.org/html/content/facts_renew_nrg.html>

People's Daily, 24 November 2000 <fpeng.people daily.com.cn/200011/24/eng20001124_56041.html>

<www.inshp.org/cover.asp>

The power of water
Energy Balances of Organization for Economic Cooperation and Development (OECD) Countries, 1960–1997, and Energy Balances of Non-OECD Countries, 1971–1997, (OECD, Paris, 1999), compiled by UNEP GEO Data Portal

44–45 THE DAMNED
Dams and their consequences
Dams and Development: A New Framework for Decision-Making. The Report of the World Commission on Dams. Earthscan, London, 2000

World Dams
World Commission on Dams

Three Gorges Dam
Robert Benewick & Stephanie Donald *The State of China Atlas*. Penguin Reference, New York and London, 1999, updated 2003 from: <www.chinaonline.com> and from International River Network <www.irn.org/programs/threeg/>

Part 3: Water Health

48–49 ACCESS TO WATER
Improved water source
The State of the World's Children 2004. UNICEF, New York, 2003

Water source
WHO/UNICEF Joint Monitoring Programme, 2002, cited in World Water Assessment Programme *Water for People: Water for Life*. UNESCO and Berghahn Books 2003 p109

Urban and rural facilities
Global Water Supply and Sanitation Assessment 2000 Report. WHO and UNICF, 2000 <www.who.int/docstore/water_sanitation_health/Globassessment/GlobalTOC.htm>

50–51 SANITATION
Adequate sanitation
The State of the World's Children 2004. UNICEF, New York, 2003

Types of sanitation
Global water supply and sanitation Assessment 2000 Report. WHO and UNICEF, 2000 <www.who.int/docstore/water_sanitation_health/ Globassessment/Global1.htm>

Wastewater treatment
World Development Indicators 2003. World Bank, New York, 2003, Indicator 3.11

52–53 DIRTY WATER KILLS
The cycle of disease
Annette Pruss *et al* 'Estimating the Burden of Disease from Water, Sanitation, and Hygiene at a Global Level' *Environmental Health Perspectives.* v.110, no. 5, May 2002

80% of illnesses
'Lessons Learned in Water, Sanitation and Health'. 1993, cited on <www.water.org/crisis/diesase.html> [*sic*]

Major water-borne diseases
WHO fact sheets on water-related disease: <www.who.int/water_sanitation_health/diseases/en/>

Deaths from dirty water
Trachoma
The World Health Report 2002. WHO, Geneva 2003, Annex Table 11

54–55 HARBOURING DISEASE
WHO fact sheets on water-related diseases: <www.who.int/water_sanitation_health/diseases/en/>

World Water Assessment Programme *Water for People, Water for Life.* WWAP, UNESCO and Berghahn Books 2003

Malaria
Dengue
The World Health Report 2003. WHO, 2003, Annex Table 2
WHO fact sheets on water-related diseases <www.who.int/water_sanitation_health/diseases/en/>

Blighted lives
The World Health Report 2003. WHO 2003, Annex Table 3

West Nile Virus in the USA
West Nile Virus on the increase
<www.cdc.gov/ncidod/dvbid/westnile/surv&contr ol03Maps_PrinterFriendly.htm>

56–57 INSIDIOUS CONTAMINATION
Payal Sampat, 'Groundwater Shock', *World Watch.* January/February 2000

Payal Sampat, 'Deep Trouble: The Hidden Threat of Groundwater Pollution'. WorldWatch Paper 154, December 2000

<ionex <www.ionex.co.uk/health.htm>

Arsenic
Mass poisoning in Bangladesh
Arun B. Mukherjee and Prosun Bhattacharya 'Arsenic in groundwater in the

Bengal Delta Plain: slow poisoning in Bangladesh' <geosci.uchicago.edu/~archer /EnvChem/> published on NRC Research Press, 2001

WHO fact sheets <www.who.int/water_sanitation_health/diseases/en/>

Fluoride
<educationa.vsnl.com/fluorosis/global/html>

WHO fact sheets <www.who.int/water_sanitation_health/diseases/en/>

Lead
CSE *The Citizen's Fifth Report.* Delhi 1999, p48, cited in World Water Assessment Programme *Water for People, Water for Life.* UNESCO and Berghahn Books 2003

Part 4: Reshaping the Natural World

60–61 DIVERTING THE FLOW
Nepal: Melamchi project
<www.adb.org/Documents/RRPs/NEP/rrp-31624-nep.pdf>

<www.adb.org/Documents/News/2000/nr2000167.asp>

<www.adb.org/Documents/News/2000/pi2000167.asp>

Spain: Proposed diversion of the Ebro
<gasa.dcea.fct.unl.pt/ecoman/projects/adviesor/publication/WP1es.pdf>

<www.unesco.org/courier/2000_12/uk/planet.htm>

China: Water for the north
<www.water-technology.net/project_printable.asp?ProjectID=2658>

<www.water-technology.net/projects/shanxi/index.html>

<www.china.org.cn/english/China/46341.htm>

River fragmentation
World Resources Institute, *Watersheds of the World.* 14 Degree of river fragmentation and flow regulation <www.iucn.org/themes/wani/eatlas/>

62–63 DRAINING WETLANDS
Protecting wetlands
EarthTrends Data Tables: Biodiversity and Protected Areas 2003 on <earthtrends.wri.org/datatables>

Threatened freshwater fish
Groombridge and Jenkins 2002; <www.redlist.org> cited in World Water Assessment Programme *Water for People, Water for Life.* UNESCO and Berghahn Books 2003, Table 6.6, p141

Garden of Eden
UNEP Vital Water Graphics <www.unep.org/vitalwater/26.htm>

Oasis of Azraq
Maude Barlow and Tony Clarke *Blue Gold.* New Press, New York and Earthscan, London 2002 p21

<www.globeandmail.com/series/deathwish/0604.html>

64–65 GROUNDWATER MINING
Payal Sampat 'Groundwater Shock', *World Watch.* Jan/Feb 2000

Payal Sampat 'Groundwater Mining'. WorldWatch Institute, 2000

Where groundwater is found
Groundwater Resources of the World, Bundesanstalt Fur Geowissenschaften und Rohstoffe <www.bgr.de/b1hydro/fachbeitraege/a200401/folie0.pdf>

Text on map:
<lanic.utexas.edu/la/Mexico/water/book.html>

World Global Trends <www.earth-policy.org/updates/update22.htm>

Payal Sampat 'Groundwater Mining'. WorldWatch Institute, 2000

'Groundwater: A North American Resource, A Discussion Paper', Expert Workshop on Freshwater in North America, 2002, prepared by Joanna Kidd, Lura Consulting <www.cec.org/files/pdf/LAWPOLICY/water_disucssion-e1.pdf>

Maude Barlow and Tony Clarke *Blue Gold.* New Press, New York and Earthscan, London 2002

66–67 EXPANDING CITIES
UN-HABITAT <www.unchs.org/articles/global_water_global%20water_1.asp>

<www.itt.com/waterbook/mega_cities.asp>

Urban water safety
Wasted water
Global Water Supply and Sanitation Assessment 2000 Report. WHO and UNICF, 2000 <www.who.int/docstore/water_sanitation_health/Globassessment/GlobalTOC.htm>

City living
World Water Assessment Programme *Water for People, Water for Life.* UNESCO and Berghahn Books 2003

World Urbanization Prospects: The 2001 Revision. United Nations Population Division, New York, 2002 <www.un.org/esa/population/publications/wup2001/wup2001dh.pdf>

City growth in the USA
United States General Accounting Office, *Water Quality GAO-01-679*, June 2001, citing US Department of Agriculture

Mexico City
David C Sommer on: <ga-mac.uncc.edu/faculty/haas/geol3190/termpap/sommer/history.html>

Christopher A Scott *et al*, 'Urban-Wastewater Reuse for Crop Production in the Water-Short Guanajuato River Basin, Mexico' International Water Management Institute Research Report 41, 2000

<www.iwmi.cgiar.org/pubs/pub041/Report41.pdf> *Mexico City's Water Supply.* National Academy Press, Washington DC, 1995

<lanic.utexas.edu/la/Mexico/water/book.html>

<www.un.org/Pubs/CyberSchoolBus/habitat/profiles/mexico.asp>

Global Water Supply and Sanitation Assessment 2000 Report. WHO and UNICF, 2000 <www.who.int/docstore/water_sanitation_health/Globassessment/GlobalTOC.htm>

68–69 DESPERATE MEASURES
Desalination
Peter H. Gleick, Pacific Institute <www.worldwater.org/table16.html>

Imported water
Maude Barlow and Tony Clarke *Blue Gold*. New Press, New York and Earthscan, London 2002

Sourcebook of Alternative Technologies for Freshwater Augmentation in Small Islands <www.unep.or.jp/ietc/Publications/TechPublications/TechPub-8d/importation.asp>

<www.tve.org/ho/doc.cfm?aid=537>

Tony Falkland, Mark Overmars, David Scott,

'Pacific Dialogue on Water and Climate'. South Pacific Applied Geoscience Commission, October 2002

Power for desalination
World Water Assessment Programme *Water for People, Water for Life.* UNESCO and Berghahn Books 2003, p89

UK desalination
Independent on Sunday quoting BBC Countryfile, 11/11/03 cited on <www.rivernet.org/prs03_05.htm#111103>

70–71 FLOODS
China: <news.bbc.co.uk/1/hi/world/asia-pacific/413717.stm>

<www.tew.org/tibet2003/t2003.resource.ext.html>

Missisissppi: <www.pbs.org/wgbh/amex/flood/maps/>

Janet Abramovitz 'Unnatural Disasters'. WorldWatch Paper 158, October 2001

Floods and landslides
<www.reliefweb.int/w/map.nsf/Country?OpenForm&Start=1&Count=1000&Expand=22&Query=Wd_World%7EwByCKeyword&Seq=1>

Rising floods
Death and disaster
International Federation of the Red Cross <www.cred.be/emdat/sumdata/wdr/wdr2002.htm#Tables%201%20to%2013>

Flooding in Bangladesh
Government of Bangladesh map, available on ReliefWeb:

<www.irn.org/pubs/wp/bangladesh.html>

72–73 DROUGHTS
Janet Abramovitz 'Unnatural Disasters'. WorldWatch Paper 158, October 2001

<www.nws.noaa.gov/om/drought.htm>

Aridity zones
UNEP/GRID 1991 <biodiv.wri.org/pubs_content_text.cfm?ContentID=722>

Life-threatening droughts
<www.reliefweb.int/w/rwb.nsf>

Greenhouse gas
World Bank *World Development Indicators 2002.*
World Bank, Washington DC, 2002, Indicator 3.8

Burden of death
International Federation of the Red Cross

<www.cred.be/emdat/sumdata/wdr/wdr2002.htm
#Tables%201%20to%2013>

Drought and desertification
UN Convention to Combat Desertification
<www.unccd.int/main.php>

Part 5: Water Conflicts
76–77 THE NEED FOR CO-OPERATION
The UN World Water Development Report, facts
and figures <www.unesco.org/water/wwap/facts_
figures/sharing_waters.shtml>

Water dependency
FAO Aquastat 2003
<www.fao.org/ag/agl/aglw/aquastat/main/>

Co-operation
Meredith A. Giordano and Aaron T. Wolf 'The
World's International Freshwater Agreements:
Historical Developments and Future
Opportunities' *Atlas of International Freshwater
Agreements.* UNEP and FAO, 2002
Water issues
Conflict and co-operation
Aaron T. Wolf, Shira B Yoffe and Mark Giordano
*International Waters: Indicators for Identifying
Basins at Risk* <webworld.unesco.org/water/
wwap/pccp/cd/pdf/water_conflict_indicators/inter_wate
rs_indicators_for_identifying_basin_at_risk.pdf>

78–79 PRESSURE POINTS
Central Asia
United Nations Environment Programme:
<www.grida.no/enrin/graphics.cfm?data_id=2328
7&country=centralasia>

<www.reliefweb.int/w/rwb.nsf/0/4b673e5e23bf8d
04c1256d6c002eb129?OpenDocument>

Alexei Kalmykov, Reuters News Service, 14
October 2003, available on:
<www.rivernet.org/prs03_04.htm#011003>

USA and Mexico
Arizona Department of Water Resources:
<www.water.az.gov/adwr/Content/Publications/fi
les/colorivmgt.pdf>

Nigel Hunt, Reuters News Service, 1 October
2003, cited on <www.rivernet.org/prs03_04.htm>

Israel and Palestine
Palestinian Academic Society for the Study of
International Affairs, Jerusalem <www.passia.org>

www.ecoisp.com/flashpoints13.asp

<www.israel-mfa.gov.il/mefa/go.asp>

'Evaluating Water Balances in Israel'
<web.idrc.ca>

<www.wws.princeton.edu>

<waternet.rug.ac.be/waterpolicy.htm>

PASSIA, citing Jad Ishaq, The Palestinian Water
Crisis, Center for Policy Analysis on Palestine,
Washington DC, 18 August 1999; B'Tselem,
Thirsty for a Solution, July 2000)

80–81 WEAPON OF WAR
Water as weapon
Peter H Gleick, Pacific Institute *Water Conflict
Chronology* <www.worldwater.org/conflict.htm>

Part 6: Ways Forward

84–85 THE WATER BUSINESS
Jorge Cuba 'Free or foreign: the water battle in Bolivia' <www.unesco.org/courier/2000_12/uk/planet2.htm>

Water Subsidies and the Environment, OECD/GD(97)220, 12E77396 <www.olis.oecd.org>

Bottled water
<www.bottledwaterweb.com/news/nw_102902.html>

<www.bottledwaterweb.com/news/nw_070201.html>

<www.nacsonline.com>

<www.foodnavigator.com/news/news-NG.asp?id=47485#>

Wages for water
cited in 'Earth', special issue published by *The Guardian*, London, August 2002

Who pays what?
Watertech Online 2001, cited in World Water Assessment Programme *Water for People, Water for Life*. UNESCO and Berghahn Books, 2003, p338

Water globalization
<www.suez.com>

The poor pay more
Asian Development Bank, *Water for All*. 1997, cited in WWAP-UNEP *Water for People, Water for Life*. UNESCO and Berghahn Books, 2003, p341

86–87 CONSERVING SUPPLIES
Industry
Estimated use of water in the United States in 1990: Trends in Water Use 1950–1990

<water.usgs.gov/watuse/wutrends.html#TAB31>

Domestic
Population Report: Solutions for a Water-Short World. John Hopkins School for Public Health <www.infoforhealth.org/pr/m14edsum.shtml>

Pacific Institute press release, posted 18 November 2003

Wastewater reuse
Christopher A Scott *et al*, 'Urban-Wastewater Reuse for Crop Production in the Water-Short Guanajuato River Basin, Mexico' International Water Management Institute Research Report 41, 2000

<www.iwmi.cgiar.org/pubs/pub041/Report41.pdf>

Population Report: Solutions for a Water-Short World. John Hopkins School for Public Health <www.infoforhealth.org/pr/m14edsum.shtml>

Agriculture
Methods of irrigation
Population Report: Solutions for a Water-Short World. John Hopkins School for Public Health <www.infoforhealth.org/pr/m14edsum.shtml>

'Success stories in water conservation in the Mediterranean region', GRID Issue 16, August 2000 <www.hrwallingford.co.uk/projects/IPTRID/grid/pdf-files/grid16articles/G16pg4b-6a.pdf>

Water harvesting
<www.humanscapeindia.net/humanscape/new/mar02/theriver.htm>

World Water Vision. Earthscan, London 2000, Chapter 5

Less water, more rice
Xijin Wu 'Development of water-saving irrigation technique on a large paddy rice area in Guangxi region of China' <www.icid.org/wat_xijin.pdf>

88–89 SETTING PRIORITIES
CEH Wallingford (formerly the Institute of Hydrology) <www.nwl.ac.uk/research/WPI/>

'Using the Water Poverty Index to monitor progress in the water sector', Centre for Ecology & Hydrology (CEH); Department For International Development (DFID), <www.nwl.ac.uk/research/WPI/images/wpileaflet a4.pdf>

Water Poverty Index
C. A. Sullivan, J. R. Meigh, A. M. Giacomello, T. Fediw, P Lawrence, M. Samad, S. Mlote, C.

Hutton, J. A. Allan, R. e. Schulze, D. J. M. Diamini, W. Cosgrove, J. Delli Priscoli, P. H. Gleick, I. Smout, J. Cobbing, R. Calow, C. Hunt, A. Hussain, M. C. Acreman, J. King, S. Malomo, E. L. Tate, D. O'Regan, S. Milner and I. Steyl 'The Water Poverty Index: Development and application at the community scale' *Natural Resources Forum*, 2003, 27, 1-11

Country profiles
Peter Lawrence, Jeremy Meigh and Caroline Sullivan, 'The Water Poverty Index: an International Comparison', Keele Economics Research Papers 2002/19 <www.keele.ac.uk/depts /ec/web/wpapers/kerp0219.pdf>

A new well in Malica
WaterAid

90–91 VISION OF THE FUTURE
<www.wordspy.com/words/virtualwater.asp>
<www.unesco.org/courier/1999_02/uk/dossier/intro31.htm>

<www.ihe.nl/vmp/articles/Projects/PRO-Virtual_Water_Trade.html>

Business as usual
Business nearly as usual
A changed world
World Water Vision. Earthscan, London 2000
<www.worldwatercouncil.org/vision.shtml>
Chapter 4 Our Vision of Water and Life in 2025

'Early on an April morning' IUCN (World Conservation Union) 1999, cited in *World Water Vision*